Pulling G

Human Responses to High and Low Gravity

Other Springer-Praxis books of related interest by Erik Seedhouse

Erik Seedhouse

Pulling G

Human Responses to High and Low Gravity

 Springer

Published in association with
Praxis Publishing
Chichester, UK

 PRAXIS

Dr Erik Seedhouse, M.Med.Sc., Ph.D., FBIS
Milton
Ontario
Canada

SPRINGER–PRAXIS BOOKS IN POPULAR SCIENCE
SUBJECT *ADVISORY EDITOR*: Stephen Webb, B.Sc., Ph.D., M.Inst.P., C.Phys.

ISBN 978-1-4614-3029-2   ISBN 978-1-4614-3030-8 (eBook)
DOI 10.1007/ 978-1-4614-3030-8
Springer New York Heidelberg Dordrecht London

Library of Congress Control Number: 2012932172

Cover design: Jim Wilkie
Photo credit: www.learning-to-fly.com
Project copy editor: Christine Cressy
Typesetting: BookEns, Royston, Herts., UK

Printed on acid-free paper

Springer is part of Springer Science+Business Media (www.springer.com)

Contents

Preface

Deeper into the atmosphere the G-forces increased to the maximum of 2. In any other circumstances this would have been a trivial force. A modern fighter can subject its pilot to 9 Gs. But for an astronaut returning from days of weightlessness, the feel of the G-forces was significantly amplified. It seemed as if an elephant were on my shoulders. I was being crushed into my seat. The weight of my helmet made it difficult for me to hold up my head. My vision began to tunnel, as if I were looking through a straw. I knew from my fighter jet experiences tunnel vision was an indication of approaching blackout. The vision area of my brain wasn't getting enough oxygenated blood.

Extract from *Riding Rockets*, by astronaut, Mike Mullane

Ever wondered what it feels like to pull 7 G in a fighter plane or what it's like to get up close and personal with a Formula One (F1) car or other gravity-defying vehicles? Pulling G is not an easy task, as it's extremely demanding on the body. But what exactly is G? G refers to G-force – a term you've no doubt heard many times to describe the gravitational force on a person or object when it's accelerating. This G-force can refer to rocket launches, fighter aircraft, or any activity involving acceleration. But the way G affects the body is determined by several factors, including the magnitude of the G-force, the length of time the person has to withstand it, the direction from which the force is applied, and even body posture at the time of the G. For example, an instantaneous impact such as a high-speed F1 accident may prove lethal, whereas exposure to transient Gs during a rollercoaster ride will have no long-term effects.

To give you an idea of how G-forces feel, merely sitting in your chair is equivalent to 1 G. If you're lucky enough to catch a ride in a Bugatti Veyron and ask the driver to accelerate from 0 to 100 km/hr (which will take an impressive 2.4 seconds), you will experience a force of 1.18 G. If you're even luckier and manage to get a ride in an F1 car (there are some two-seaters used to entertain corporate sponsors) and ask the driver to demonstrate its performance envelope, you will experience about 1.45 G. Back in the day when the Space Shuttle was flying, astronauts were subjected to 3.2 G during launch. Take a ride on one of the big scream-machine roller coasters and you can expect to really feel your organs displace inside your ribcage. For example, the Desperado subjects riders to a downward force of negative 4.1 G in its first drop. Snapping back upward out of this drop, riders experience 2.8 G before being dropped again and slung around to face a negative 5.25-G rightward turn. In addition to the stomach-churning twists and turns, the Desperado subjects riders to nine instances of weightlessness; no other roller coaster comes close. If the Desperado doesn't satisfy your G craving, you can ramp up the G level by taking a ride in an aerobatic aircraft, which routinely pull 7 or 8 positive and negative Gs.

These are just some of the examples used in this book to describe the risks of the high and low-G environment and the physiology of surviving G. *Pulling G* begins with an explanation of biodynamics and Colonel Stapp's pioneering research before describing the phenomenon of gravity-induced loss of consciousness (G-LOC) and the body's compensatory mechanisms. In the next chapter, the dangerous issue of "push–pull" is tackled before introducing the human centrifuge and how it is used as a training tool for astronauts and fighter pilots. Next is a description of the lateral G experienced in the F1 arena and how drivers train to tolerate high G-loads at right angles to the spine. The following chapter provides an insight into the experiences of pilots who have survived high-speed ejections. After describing the accelerative forces that are unleashed when punching out from a high-performance jet, the focus is directed at the stresses experienced during launch and re-entry before segueing to the subject of zero-G and how human physiology adapts to transient microgravity. Finally, staying on the subject of space, the final chapter describes how scientists are investigating using artificial gravity to research ways of reducing the effect of zero-G on astronauts' bodies.

Acknowledgments

In writing this book, the author has been fortunate to have had five reviewers who made such positive comments concerning the content of this publication. He is also grateful to Maury Solomon at Springer and to Clive Horwood and his team at Praxis for guiding this book through the publication process. The author also gratefully acknowledges all those who gave permission to use many of the images in this book, especially AMST and Mark Holderman.

From left to right: Warrant Officer Chris Kelly, the author, MCpl Allison Riddell, and Sgt Chris Townson seated in the control room of Canada's only human centrifuge located in Downsview, Ontario. This is a rather unique photo because these are the only certified current Acceleration Training Officers in Canada. When they're not spinning pilots in the centrifuge, Chris Kelly instructs Conduct after Capture training. Chris Townson indulges his penchant for motorbikes, and Allison earns a small fortune renovating houses. Image courtesy: David Brookes

The author also expresses his deep appreciation to Christine Cressy, whose attention to detail and patience greatly facilitated the publication of this book, to Jim Wilkie for creating the cover of this book, and to Stewart Harrison who sourced several of the references that appear in this book.

Once again, no acknowledgment would be complete without special mention of our rambunctious cats, Jasper, Mini-Mach, and Lava, who provided endless welcome (and occasionally unwelcome!) distraction and entertainment.

*This book is dedicated to my very good friend Capt Daniel "BOOYA",
who assures me he can pull 2 G in his Sea King helicopter.*

About the author

Erik Seedhouse is a Norwegian–Canadian suborbital astronaut whose life-long ambition is to work in space. After completing his first degree in Sports Science at Northumbria University, the author joined the legendary 2nd Battalion the Parachute Regiment, the world's most elite airborne regiment. During his time in the "Para's", Erik spent six months in Belize, where he was trained in the art of jungle warfare. Later, he spent several months learning the intricacies of desert warfare on the Akamas Range in Cyprus. He made more than 30 jumps from a Hercules C130 aircraft, performed more than 200 abseils from a helicopter, and fired more light anti-tank weapons than he cares to remember!

Upon returning to the comparatively mundane world of academia, the author embarked upon a master's degree in Medical Science at Sheffield University. He supported his studies by winning prize money in 100-km running races. After placing third in the World 100-km Championships in 1992 and setting the North American 100-km record, the author turned to ultradistance triathlon, winning the World Endurance Triathlon Championships in 1995 and 1996. For good measure, he also won the inaugural World Double Ironman Championships in 1995 and the infamous Decatriathlon, an event requiring competitors to swim 38 km, cycle 1,800 km, and run 422 km. Non-stop!

Returning to academia in 1996, Erik pursued his Ph.D. at the German Space Agency's Institute for Space Medicine. While conducting his Ph.D. studies, he still found time to win Ultraman Hawaii and the European Ultraman Championships as well as completing the Race Across America bike race. Due to his success as the world's leading ultradistance triathlete, Erik was featured in dozens of magazines and television interviews. In 1997, *GQ* magazine nominated him as the "Fittest Man in the World".

In 1999, Erik decided it was time to get a real job. He retired from being a professional triathlete and started his post-doctoral studies at Vancouver's Simon Fraser University's School of Kinesiology. In 2005, the author worked as an astronaut training consultant for Bigelow Aerospace and wrote *Tourists in Space*, a training manual for spaceflight participants. He is a Fellow of the British Interplanetary Society and a member of the Space Medical Association. Recently, he was one of the final 30 candidates in the Canadian Space Agency's Astronaut Recruitment Campaign. Erik works as a manned spaceflight consultant, professional speaker, triathlon coach, and author. He is the Training Director for Astronauts for Hire (www.astronauts4hire.org) and completed his suborbital astronaut training in May 2011. He is also Canada's only High Risk Acceleration

Training Officer, which is a long-winded way of saying he spins people in Canada's centrifuge.

In addition to being a suborbital astronaut, triathlete, sky-diver, pilot, and author, Erik is an avid mountaineer and is currently pursuing his goal of climbing the Seven Summits. *Pulling G* is his tenth book. When not writing, he spends as much time as possible in Kona on the Big Island of Hawaii and at his real home in Sandefjord, Norway. Erik and his wife, Doina, are owned by three rambunctious cats – Jasper, Mini-Mach, and Lava – none of whom has expressed any desire to travel into space but who nevertheless provided invaluable assistance in writing this book (!).

Figures

Tables

Abbreviations and acronyms

ACES	Advanced Concept Ejection Seat
ACM	Aerial Combat Maneuver
ADR	Accident Data Recorder
AGSM	Anti-G Straining Maneuver
AML	Aeromedical Laboratory
AMPDXA	Advanced Multiple-Projection Dual-energy Absorptiometry
ARF	Accident Reconstruction Facility
ATO	Acceleration Training Officer
BMD	Bone Mineral Density
CAM	Centrifuge Accommodation Module
CAP	Coriolis Acceleration Platform
CF	Canadian Forces
CG	Center of Gravity
COLBERT	Combined Operational Load Bearing External Resistance Treadmill
COMBAT EDGE	Combined Advanced Technology Enhanced Design G-Ensemble
CST	Crew Space Transportation
DART	Directional Automatic Realignment of Trajectory
DCIEM	Defence and Civil Institute of Environmental Medicine
DoF	Degrees of Freedom
DRDC-T	Defence Research Development Canada Toronto
DXA	Dual-Energy Absorptiometry
ECG	Electrocardiogram
ECLSS	Environmental Control Life Support System
EEG	Electroencephalogram
EMI	Electromagnetic Interference
EPA	Eicosapentaenoic acid
FDA	Federal Drug Administration
FIA	Federation Internationale de L'Automobile
FLSC	Flexible Linear Shaped Charge
FoV	Field of View
G-LOC	Gravity-induced Loss of Consciousness
GNS	Gender Neutral Study
GVID	G-induced Vestibular Dysfunction
HAFB	Holloman Air Force Base
HANS	Head and Neck Support System
HMS	Helmet-Mounted Sight

HUD	Heads-Up Display
IMU	Inertial Measurement Unit
JAXA	Japanese Space Agency
JSC	Johnson Space Center
JSF	Joint Strike Fighter
KEAS	Knots Equivalent Air Speed
LAS	Launch Abort System
LES	Launch Escape System
LFB	Liquid Filled Bladder
LSM	Life Support Module
MARES	Muscle Atrophy Research and Exercise System
MDC	Miniature Detonation Cord
MIT	Massachusetts Institute of Technology
MMSEV	Multi-Mission Space Exploration Vehicle
MPCV	Multi-Purpose Crew Vehicle
MRI	Magnetic Resonance Imaging
NACA	National Advisory Committee for Aeronautics
NASTAR	National Aerospace Training Center
NATO	North Atlantic Treaty Organization
NAUTILUS	Non-Atmospheric Universal Transport Intended for Lengthy United States Exploration
NBL	Neutral Buoyancy Laboratory
NVG	Night Vision Goggle
OMM	Outer Mission Module
PARS	Passive Arm Restraint System
PIRD	Powered Inertia Restraint Device
PLL	Peripheral Light Loss
PPB	Positive Pressure Breathing
PPE	Push–Pull Effect
PPG	Pressure Breathing for G
PV	Photovoltaic
RMS	Remote Manipulator System
SACM	Simulated Aerial Combat Maneuver
SRC	Short Radius Centrifuge
SRSF	Slow Release Sodium Fluoride
STP	Supersonic Transition Problem
TLSS	Tactical Life Support System
USAF	United States Air Force

1

Project MX981

The fastest man on Earth and the birth of biodynamics

It is 1946, just a few months after the end of the Second World War. A B-17 Flying Fortress arcs skyward on an urgent mission. Stripped of guns and bombsights, the four-engine bomber has modified engines allowing it to do something that no other B-17 has ever been able to do: operate in the stratosphere. It cruises for hours at altitudes of nearly 14,000 m, its flight crew shivering in biting cold while, in the rear fuselage, a solitary scientist conducts a set of risky experiments, seemingly oblivious to the double-digit sub-zero temperatures. Captain John Paul Stapp (Figure 1.1), a medical doctor, is studying the effects of high-altitude flight in an attempt to determine survival times for humans at high altitudes; could aircrew exposed to the hostile conditions of stratospheric altitudes function, physically and rationally? How could they keep themselves from freezing or succumbing to decompression sickness? These were some of the questions that Stapp sought to answer and, one by one, he did. After spending the best part of three days in the air, Stapp found the answer: if a pilot breathed pure oxygen for 30 min prior to take-off (a pre-breathe), decompression sickness (DCS) could be avoided. In the field of aerospace medicine, the pre-breathe procedure was an enormous breakthrough. Thanks to Stapp's pioneering efforts, the sky was now the limit.

The development of the pre-breathe procedure also pushed Stapp to the forefront of the Aero Medical Laboratory (AML), a facility whose mandate was to study medical and safety issues in aviation. For Stapp, the lab was a perfect match for his talents; during the Second World War, the facility had produced a steady stream of innovations including advanced breathing systems, parachutes, and pressure suits – it had also established itself as the premier facility in the world for the study of the new science of biodynamics. As a reward for his efforts in resolving the DCS problem, Stapp was assigned to another AML research project designated only by an alphanumeric code: MX981. The goal of MX981 was to study human deceleration. Simply put, this study investigated the human body's ability to withstand G-forces. According to most sources at the time, 18 G was the most a human could be exposed to and survive, which was why military cockpits were built to withstand an 18-G impact. But, during the war, there had been a number of events which cast doubt on this figure. For example, there had been documented cases in which Navy pilots had crashed into aircraft carriers at high speed. Current thinking said they should have been killed but, in many cases, the pilots had walked away. There were also a number of low-magnitude

Figure 1.1 John Paul Stapp, M.D., Ph.D., Colonel, USAF (July 11th, 1910–November 13th, 1999) was a USAF flight surgeon and pioneer in studying the effects of acceleration and deceleration forces on humans. He received a bachelor's degree in 1931 from Baylor University, Waco, Texas, a MA from Baylor in 1932, a Ph.D. in Biophysics from the University of Texas at Austin in 1940, and a MD from the University of Minnesota in 1944. He also received an honorary Doctor of Science from Baylor University. Stapp joined the Army Air Corps in 1944. Two years later, he was transferred to the Aero Medical Laboratory at Wright Field as project officer in the Biophysics Branch. His first assignment included testing oxygen systems in unpressurized aircraft to mitigate the danger of decompression sickness. Dr Stapp retired from the USAF with the rank of colonel in 1970. Image courtesy: USAF.

yet fatal crash landings in which pilots' seats broke loose or their harnesses failed. Many of the AML scientists suspected these pilots had survived the initial impact only to be killed by the structural failure of the cockpit and its affiliated components. It was a problem that could only be solved by applying the science of biodynamics.

Biodynamics is concerned with the aeromedical consequences of sustained acceleration and deceleration which produce significant physiological changes in humans. Even before Stapp and the AML, it was common knowledge that G-forces have less to do with speed than with acceleration, which is the change in speed over time. If speed alone could cause the thrill that comes from feeling G-forces, then simply driving on the highway would suffice. A common perception of acceleration is of a sports car doing 0–100 km/hr in four seconds, but acceleration is technically *any change in the velocity* of an object: speeding up, slowing down, and changing direction are all types of acceleration. That's why, on a rollercoaster such as the Millennium Force (Figure 1.2), you feel G-forces during tight bends that cause you to be thrown against the side of your seat (a change in direction) and also when you plunge from the heights (accelerate) or grind to a halt (decelerate).

Riding the Millennium Force, you feel the thrill but don't black out, because the coaster's creators designed it to be well within the G-force tolerance of the average person. The amount of G that is tolerable differs from individual to individual but, no matter who you are, G-tolerance depends on three factors: the direction in which the G-forces are felt, the number of Gs involved, and how long those Gs last. Depending how your body is oriented when it accelerates, you can feel G-forces either from front-to-back, side-to-side, head-to-toe, or toe-to-head. Most of us don't have a problem tolerating the two horizontal axes but have a harder time dealing with the vertical or head–toe axes. That's because vertical forces have a pronounced effect on blood pressure. At sea level, or 1 G, the human body needs 22 mm of mercury blood pressure to pump sufficient blood from the heart to the brain. When subjected to 2 G, we need twice that pressure, at 3 G we need three times, and so on. Most of us would lose consciousness when head-to-toe G-forces reach about 4 or 5 because our hearts just can't generate the necessary pressure to send sufficient blood to the brain. Instead, blood pools in our lower extremities and our brains fail to get enough oxygen. As we'll see in the next chapter, fighter pilots can handle greater head-to-toe G-forces (up to 8 or 9 G) by wearing anti-G suits. These specialized custom-fit outfits use air bladders to constrict the legs and abdomen during high Gs to keep blood in the upper body. Fighter pilots can also increase their G-tolerance by learning breathing and muscle-tensing techniques while training in centrifuges, which is discussed in Chapter 4.

Magnitude and duration are as critical as direction. While Stapp demonstrated that people can withstand much higher G than had long been thought, there is a limit to what a human can take. As long as the G-forces are sustained for only short periods, humans can tolerate surprisingly high Gs. For example, the average sneeze creates almost 3 G, a slap on the back about 4 and, if you decide

Figure 1.2 The Millennium Force is a steel roller coaster located at Cedar Point, Sandusky, Ohio. Standing 95 m tall, the roller coaster, which has a top speed of 150 km/hr and subjects riders to 4.5 G, has been voted the number one steel roller coaster in the world in Amusement Today's Golden Ticket awards six times since the year 2001. Image courtesy: Wikimedia Commons.

to jump from a meter up and land stiff-legged, you'll feel about 100 G momentarily. The key word is *momentarily*; we suffer no ill-effects from these everyday events because they're so brief. Having said that, G is a unique environment, since protection from its effects is impossible. While pilots can be protected from other austere environments such as high altitude (thanks to pressure suits) or extreme cold (insulated clothing), they can *never* be completely protected against G. It's why the science of biodynamics is such an important one, so it's worthwhile understanding the physical principles that are involved before returning to Stapp's story.

Biodynamics

Humans are adapted to live and survive within the ever-present, accelerative force of gravity. While on Earth, this is a constant, and we live and function with it from the day we're born until the day we die. It doesn't take us long to become accustomed to gravity at the standard 1 G but, when we pilot an aircraft, all that we've learned about gravity changes because flight is all about overcoming gravity. Getting the upper hand on gravity is achieved by acceleration, which is described in units of force called "Gs". A pilot in a steep turn may experience forces of acceleration equivalent to many times the force of gravity. This is

Figure 1.3 The F-22 Raptor is a single-seat, twin-engine fighter designed primarily as an air superiority fighter. It is super-maneuverable, capable of greater than 6-G/second onset rates and pulling more than 9 G. Image courtesy: USAF.

Figure 1.4 In the Red Bull Air Race World Championship, competitors fly through a series of air gates, between which they perform prescribed maneuvers, many of which subject the pilots to high G-loads. Image courtesy: Red Bull.

especially true in military fighter jets (Figure 1.3) and high-performance, aerobatic aircraft, which can crank out more than 9 G. Air race pilots (Figure 1.4) in a tight pylon turn also experience high G-forces.

Another important aspect of understanding biodynamics is the three types of acceleration (Figure 1.5). These types are referred to as Linear, Radial, and Angular Acceleration. Linear Acceleration reflects a change of speed in a straight line as experienced during take-off, landing, or in level flight. Radial Acceleration is the result of a change in direction such as when a pilot performs a sharp turn, pushes over into a dive, or pulls out of a dive, while Angular Acceleration results from a simultaneous change in both speed and direction, which happens in spins and climbing turns. During flight, a pilot may experience a combination of these accelerations as a result of input to the flight controls.

In Figure 1.5, Gx is the force acting on the body between chest and back. Positive Gx (+Gx) is experienced during the take-off roll as the throttle is advanced (this is the force that pushes the pilot back into the seat as the aircraft accelerates) and negative Gx (–Gx) is the force from back to chest, and is encountered during landing as the throttle is closed (this force pushes the pilot forward into the shoulder strap). Naval pilots flying from aircraft carriers (Figure 1.6) feel the extremes of +Gx during catapult launches as the aircraft accelerates to 260-plus km/hr in just under two seconds.

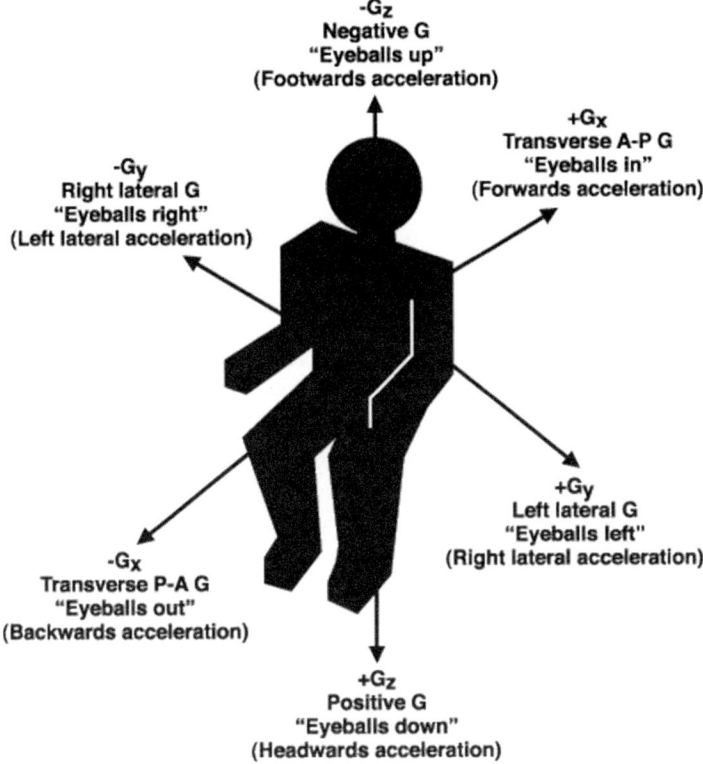

Figure 1.5 Positive G, or +Gz, acceleration occurs when the body is subject to acceleration in the head-ward direction. The inertial force acts in the opposite direction towards the feet, and the body is forced into the cockpit seat. Negative G, or –Gz, acceleration occurs when the body is accelerated foot-ward. The inertial force is towards the head, and the body is lifted out of the cockpit seat. Forward G, or +Gx, acceleration occurs when the accelerative force acts in the chest-to-back direction. Backward G, or –Gx, acceleration occurs when the accelerative force acts in the back-to-chest direction. Right or left-lateral G, or $\pm$ Gy, acceleration occurs when the accelerative force impacts across the body from a shoulder-to-shoulder direction. Image courtesy: FAA/NASA.

The second force depicted in Figure 1.5 is *Gy*, a lateral force that acts from shoulder to shoulder, usually encountered during aileron rolls. Aerobatic pilots routinely encounter Gy but don't have any problems maneuvering their aircraft. The final force shown is *Gz*. This is applied to the vertical axis of the body and, if it is experienced from head to foot, it is termed positive Gz (+Gz). This happens when a pilot pulls out of a dive or pulls into an inside loop. The other type of Gz is negative Gz (–Gz), which travels from foot to head. It's experienced when a pilot pushes over into a dive. The forces described are referred to during the book

Figure 1.6 An aircraft catapult is a device used to launch aircraft from aircraft carriers. It consists of a track built into the flight deck, below which is a shuttle attached through the track to the nose gear of the aircraft. Image courtesy: USAF.

but, in this chapter, we'll be talking about +Gx and –Gx because Stapp's problem focused on acceleration and deceleration.

John Paul Stapp

In 1947, to help the military resolve its G-related problems, Captain Stapp traveled to Muroc Air Base in the Mojave Desert to view a device known as the "human decelerator". Muroc, which eventually became Edwards Air Force Base (EAFB), was a remote base, but it had a critical item of equipment needed to measure deceleration: a 600-m-long rocket sled track.[1] Built by Northrop engineers, the human decelerator comprised a sled and a track that was basically just a standard railroad rail set in concrete. The sled, built out of welded tubes and designed to withstand 100 G, was dubbed the "Gee Whiz" (Figure 1.7).

---

[1] The Germans had originally built a special track and rocket-powered test vehicle. After the Germans' defeat in the Second World War, the apparatus had ended up with the Soviets and Stapp was sent to Moscow to examine it. He pronounced the device "ingenious".

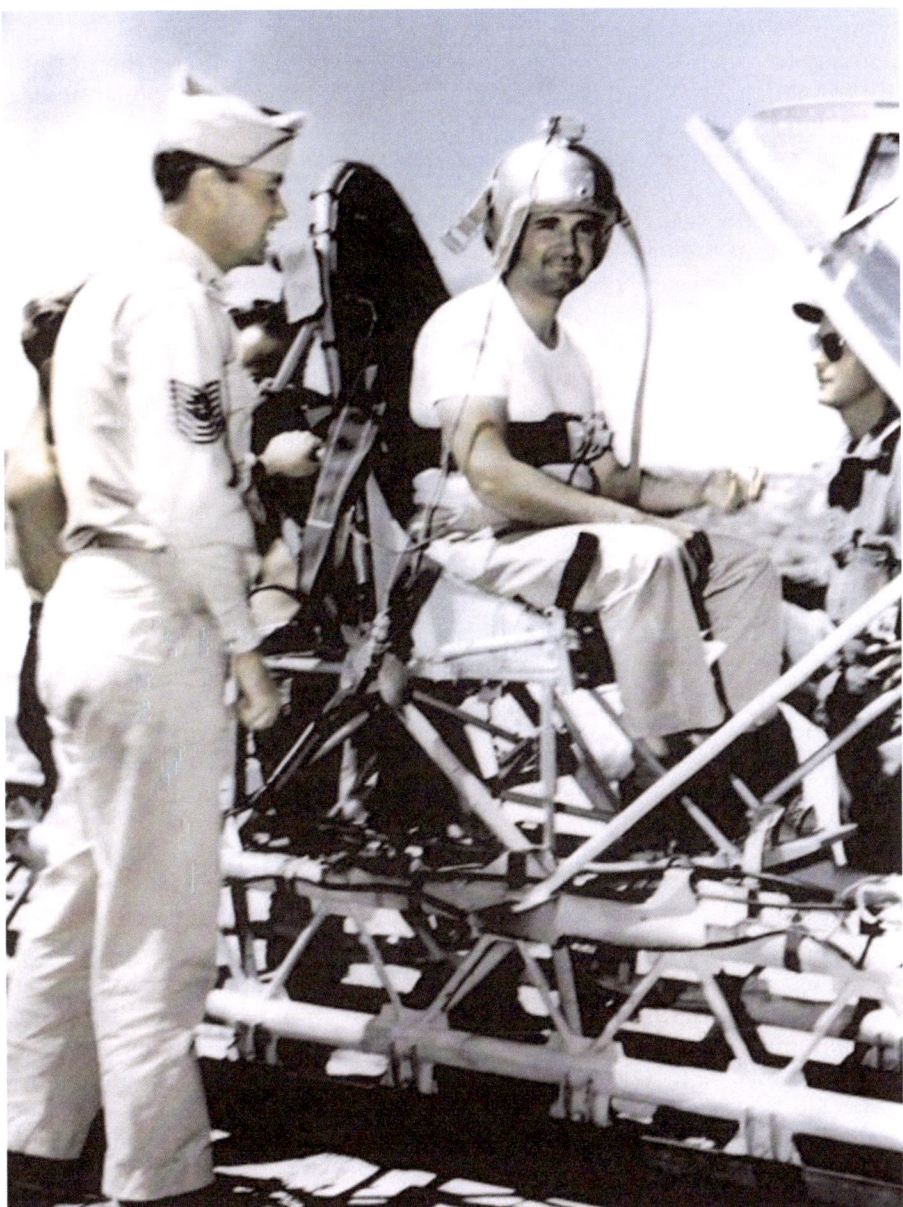

Figure 1.7 The "human decelerator" (the "Gee Whiz") consisted of a 680-kg carriage mounted on a 610-m railroad track supported on a heavy concrete bed and a 14-m hydraulic braking system. Four slippers secured the carriage to the rails while permitting it to slide freely. At the rear of the carriage, rockets provided the propelling force. Braking was accomplished by partitioned scoops that picked up the water and threw it forward. Image courtesy: Edwards Air Force Base.

About 5 m long and 2 m wide, the "Whiz", which weighed nearly 700 kg, was designed to rocket along the track at up to 250 km/hr. Stopping the sled at the end of the track was a set of 13-m-long hydraulic brakes that bore an uncanny resemblance to dinosaur teeth. The brakes were capable of slowing the rocket sled to half its speed in one-fifth of a second, thereby producing G-forces equivalent to those generated in an aircraft crash.

On top of the sled chassis was a lightweight metal cab that housed a rugged, custom seat. At the back were a telemetry antenna mast and a rack designed to hold up to four rocket bottles. The bottles, the same type as that used to boost heavy aircraft off short runways, were each capable of generating 5,000 pounds of thrust. By varying the number of bottles and the brake pressure, a range of G-forces could be applied to the sled and its occupant – which brings us to the unusual character of Oscar Eightball.

Oscar was fearless, incredibly tough, and completely brain-dead, so it wasn't surprising that nearly everyone involved in the alphanumerically designated project agreed that Oscar was the perfect test subject for the life-threatening tasks at hand. After all, if something terrible did happen, Oscar could simply be replaced because Oscar was a dummy. But Stapp had other ideas. He didn't warm to the idea of letting Oscar do the tests, deciding that the only way to truly experience the effects of deceleration was to experience them at first hand. The other experts at the test site thought Stapp had taken leave of his senses and brought in an expert from the Massachusetts Institute of Technology (MIT), who promptly pronounced that anyone subjected to more than 18 G would break every bone in their body. Such prophecies of doom had no effect on Stapp who, using a slide rule and his knowledge of human anatomy, concluded the 18-G limit was nonsense. Instead, he predicted that the human limit probably lay closer to 40 G. To the engineers and the expert from MIT, this was crazy talk. But, back in those days, pioneers had a habit of proving the experts wrong. For example, just a few months earlier, Chuck Yeager (for whom Stapp had served as flight surgeon), flying in the Bell X-1 (Figure 1.8) rocket plane, broke the sound barrier and, in doing so, defied all the doomsayers, some of whom had predicted his body would turn into tapioca pudding and he would lose his ability to speak.

Stapp might have been a maverick, but he wasn't crazy, so, when it came to the initial tests, he let Oscar be the guinea pig. Just in case. It proved to be a wise move. On the first run, on April 30th, 1947, the hydraulic brakes and backup restraint system failed, and the sled slid off the track and into the desert. It wasn't badly damaged, but the brakes were almost written off; a series of steel teeth designed to trip cams had broken clean off on impact. Another test featured Oscar rocketing along the track at 250 km/hr wearing only a light safety belt. At the end of the run, the brakes locked up, instantly producing 30 G. Predictably, the belt snapped and Oscar went sailing right through an inch-thick wooden windscreen as if it were paper, finally coming to a halt more than 200 m downrange. Stapp was undeterred however, and pressed on with more test runs until he felt confident in the system. Some of these test runs involved animals. The animal runs were invaluable to the scientists, as they demonstrated what

Figure 1.8 The Bell X-1 was a joint NACA–US Army/US Air Force supersonic research project. It reached nearly 1,600 km/hr in 1948. A derivative of this same design, the Bell X-1A, was the first aircraft to exceed the speed of sound in controlled, level flight, and was the first of the so-called X-planes. Image courtesy: USAF.

humans might safely do and, equally importantly, what they shouldn't attempt. For example, early in the test-and-verification phase of the program, someone suggested that the pilot of a high-speed aircraft might be safer if the ejection seat blew him into the air blast head-first. The hypothesis was duly tested by strapping an anesthetized chimpanzee on the sled head-forward and blasting the unsuspecting animal down the track before bringing the sled to a 270-G stop. It was presumed the chimp died instantly, although it was impossible to tell because there wasn't much left of the animal. Not surprisingly, there was no second test! The chimpanzee wasn't the only animal to be sacrificed in the name of biodynamics, since animals rode the sled more often than Stapp. Many experienced much higher G-forces and, in some cases, were simply vaporized by the G-forces, the wind blast, or a combination of both. In overall proportions and in some details of internal structure, the chimpanzee is actually quite similar to man but, in certain aspects of spinal structure, the bear seemed a better fit. At the time, bears had the added advantage of being cheaper and more plentiful than chimpanzees, although not as abundant as hogs. Hogs, while plentiful, at least among the large animals, also had their points of resemblance to the human body. Since they were also the most edible, the hogs made good barbecue material! When sacrificed after a test, hogs were occasionally presented to different units at Holloman, including the Aeromedical Field Laboratory itself, for use in group picnics, but this was only incidental to their primary research function.

Finally, in December 1947, after eight months, 35 test runs, and more than a few disintegrated chimps, fragmented hogs, and dismembered bears, Stapp was ready to attempt a manned run. Erring on the side of caution, Stapp used only one rocket for the first run and faced backwards to minimize the effects of the G-load. During the first run, the sled reached about 140 km/hr and Stapp experienced a deceleration of only 10 G. Emboldened by such an easy ride, Stapp immediately added two more rockets. This time, the sled reached 320 km/hr! Over the next few weeks, Stapp experimented with the number of rockets used with various braking configurations to vary the rate of onset (the time it took for forces to build to a maximum). By August of the following year, Stapp had completed 16 runs and had survived a jaw-dropping 35 G, but not without the occasional incident and injury. As he increased the rate of acceleration and deceleration (Figure 1.9), he suffered cracked ribs, lost dental fillings, suffered the occasional concussion, and even broke his wrist (it was just as well he was a physician!). But, of all the G-inflicted trauma, the one that most concerned Stapp was the effect on vision. When facing backwards during decelerations of 18 G, Stapp experienced blurry vision as a result of blood rapidly leaving his eyeballs and pooling towards the back of his head. Clearly, when it came to G-forces, the most vulnerable part of human anatomy was the eyes.

Despite being battered and bruised by the tests, Stapp refused to allow anyone else to ride the rocket sled. This was hardly surprising, since, whether Stapp showed fear or not, the tests were terrifying (he would often throw up before each run) and he must have been loath to expose anyone else to the dangers. The track was akin a giant gun barrel and Stapp was riding the bullet. No less a man than Air Force pilot Joseph Kittinger,[2] who worked with Stapp later in his career on Project Man High, called Stapp the bravest man he ever met – quite a compliment coming from one of the most fearless men ever to walk on Earth! Eventually, Stapp relented and allowed volunteers to make some runs, although, whenever there was a new profile to be developed, it was Stapp (or some unsuspecting chimp) who was the first test subject.

After several months of tests, Stapp presented his work to higher-ranking AML officers. Predictably, they were appalled that a military officer had taken such extreme risks but, because Stapp had shown such initiative, they decided to reward him by promoting him to the rank of Major. He was also ordered to

---

[2]   It was during Project Man High that Joe Kittinger set the altitude record for parachuting by jumping from a balloon at 31,300 m.

Figure 1.9 Before the Second World War, aviation experts agreed the maximum G-force a human body could withstand was 18 G, so cockpits were built to cope with 18 G of stress. But, during the Second World War, there were many instances of pilots surviving extremely high-impact crashes; one Navy pilot survived the wreckage of his Grumman F4F Wildcat that he'd just flown into the side of its native aircraft carrier, generating an estimated 30 G. After the war, Captain John Stapp was sent to investigate. He asked for access to the rocket sled track at Edwards Air Force Base, known as the "Gee Whizz". After a couple of test runs with dummies, of which two out of three ended in catastrophic failure, Stapp decided to try it out for himself. His first run at 145 km/hr generated 10 G, so the next day, he doubled the speed. Within two months, he had completed multiple runs at the supposedly lethal 18-G barrier so he decided to double that, too. Stapp experienced 35 G. Despite eyeballs that hemorrhaged and cracked ribs, he wanted to go faster. Seven years later, he rode the Sonic Wind rocket sled which topped out at 1,011 km/hr and decelerated to zero in less than two seconds, subjecting Stapp to 40 G. During both the acceleration and deceleration, his body was never under less than 25 G, which is the equivalent of hitting a brick wall at over 120 mph. Quite the conversation starter! Image courtesy: Edwards Air Force Base.

discontinue human testing and to use chimpanzees instead. Unfazed, Major Stapp returned to Muroc, convinced his superiors would see sense once they had reviewed the data. He was right. There was no denying that Stapp's data were having an impact, since the rocket sled had proven just how ineffective some of the aircraft restraint systems were and how rear-facing passengers could survive higher G-loads than forward-facing passengers. Stapp had also discredited the 18-G limit by proving that humans could survive G-loadings in excess of 30 G. So the rocket sled tests continued until June 1951. By that time, Stapp had survived a 46-G (the highest G-force voluntarily experienced by a human) run with an onset rate of 500 G/second and a 38-G run with an onset of nearly 1,300 G. He might have attempted higher G-loadings but he had reached the design limit of the sled and its brakes. By the time Stapp had wrapped up his rocket sled tests, 74 manned runs had been made and 80 additional runs had been conducted using chimpanzees. The outcome of these tests was the adoption of a standard 32-G strength requirement for aircraft seats, the development of a new pilot's harness, and new passenger restraints.

While the rocket tests hadn't allowed Stapp to answer all the crash deceleration questions, there were new issues that had emerged. For example, in 1951, no pilot had ever ejected from an aircraft at supersonic speed and lived to tell about it and very little was known about the effects of wind blast and deceleration on a pilot ejecting at those speeds. The high-speed ejection issue was one of imposing magnitude. A pilot ejecting at transonic or supersonic speed had to first face the ejection force (see Chapter 6) required to launch him out of his aircraft followed by the sudden windblast and wind-drag deceleration, which was likely to be followed by dangerous tumbling and spinning. Any one of these forces taken separately was a potential cause of injury or death, not to mention the anxiety on the part of the pilots who didn't know whether they would survive. Back in the 1950s, the few escape systems that existed were either woefully inadequate or were simply unproven above Mach 1 and 14,000-m altitude. But, since aircraft with this range of performance already existed (Figure 1.10), there was a need for data on human tolerance to all the forces a pilot might be subjected to during such an ejection event.

So, to remedy the situation, Test Directive 5200-H1 for Biophysics of Abrupt Deceleration, dated 15th April, 1953, set forth the following objective:

> "A program of experiments with the High Performance Linear Decelerator to study tolerance and survival limits for (1) Linear Deceleration, (2) Windblast in a Linear Deceleration Field, (3) Tumbling in a Linear Deceleration Field, and (4) Linear Deceleration with Tumbling and Windblast, as factors of the problem of escape from high-speed, high altitude aircraft. Recommended limiting values established by these experiments will determine the design of escape devices and the choice of ejection seats or of ejection capsules for a particular aircraft."

The test directive added that the "current military need" was to study tolerance

Figure 1.10 The McDonnell F-101 Voodoo was a supersonic military jet fighter which set a world speed record of 1,943.4 km/hr on December 12th, 1957. Image courtesy: USAF.

to deceleration up to 55 G, although this figure was subsequently revised. In any case, the maximum number of Gs was only one of the factors to be studied. Not only were tumbling and windblast to be explored, but the rate of onset and duration of G-forces would also be considered.

Before long, the Biophysics of Abrupt Deceleration project was transformed and broadened into Project 7850, Biodynamics of Human Factors in Aviation – a move that was sought by Stapp as a means of clearly establishing the independence of his own research. It also made specific provision for research on certain topics not covered by his original test directive such as investigating tolerance to impact forces, tolerance to pressure changes, tolerance to windblast, and tolerance to aircraft crash forces.

To answer those questions, it became clear that another, more robust and powerful rocket sled would have to be built. So, in 1953, Stapp relocated to the Aeromedical Field Laboratory at Holloman Air Force Base (HAFB) in New Mexico, where there was a 1,075-m-long sled track, originally built to test the Snark missile.[3] The track terminated in a segment that could be dammed and filled

---

[3]   The Northrop SM-62 Snark was a specialized intercontinental cruise missile fitted with a nuclear warhead operated by the US Strategic Air Command from 1958 until 1961. Named after Lewis Carroll's fictional animal species, the Snark was developed to offer a nuclear deterrent to the Soviet Union at a time when ICBMs were still in development. It was the only intercontinental surface-to-surface cruise missile ever deployed by the USAF. With the deployment of ICBMs, it was rendered obsolete and taken out of service.

Figure 1.11 The Sonic Wind No. 1 was a rocket sled ridden by John Paul Stapp in the 1950s. It is currently located in the John P. Stapp Air and Space Park at the New Mexico Museum of Space History, Alamogordo, New Mexico. Image courtesy: Wikimedia Commons.

with water. By fitting a sled with water scoops and varying the water depth, different braking speeds and durations could be produced. Northrop started construction of a new sled, the Sonic Wind No. 1 (Figure 1.11), a beast of a vehicle designed to carry up to 12 rockets (Figure 1.12) and capable of producing more than 50,000 pounds of thrust.

As well as being more powerful than the Gee Whiz, the Sonic Wind No. 1 was more sophisticated, employing a two-stage design. After the rocket bottles burned out, the "propulsion sled" would be jettisoned, allowing the sled carrying the occupant to continue without the extra weight. Engineers reckoned that the Sonic Wind would be capable of shooting along the track at speeds exceeding 1,000 km/hr and withstanding up to 150 G:

> "That sled is going so damn fast the first bounce is going to be Albuquerque. I mean, there was no way on God's Earth that sled could stop at the end of the track. No way."
>
> Joe Kittinger's observations of Lt Col. Stapp's
> subsonic sled run, December 10th, 1954

Figure 1.12 Sonic Wind No. 1, the rocket sled ridden by John Paul Stapp in the 1950s, is now on display at the New Mexico Museum of Space History, Alamogordo, New Mexico. In this image, you can see the rockets that propelled the sled along the track. Image courtesy: Wikimedia Commons.

As with the Gee Whiz, the Sonic Wind was tested with a test dummy, this one dubbed Sierra Sam, a second-generation crash test dummy. After Sierra Sam survived the initial test, additional tests were made with a chimpanzee in January 1954 and then, on March 19th, Lt Col. Stapp, who had been promoted again, made his first trip down the track. Boosted by six rockets, the Sonic Wind accelerated to 673 km/hr in five seconds, and was still traveling at 500 km/hr when it hit the water brake. In the space of just 60 m, the Wind slowed to 245 km/hr, generating up to 22 G. For a moment, Stapp's body weighed more than 1,600 kg. Far from complaining about the effects of what, for most people, would have been a traumatic experience, Stapp announced he was ready to go again that same afternoon. Frustratingly for Stapp, his next sled run didn't take place for nearly five months due to problems encountered when preparing the sled to study the effects of windblast. To do this, engineers had to add a pair of doors to the sled's windscreen. The doors were designed to be tripped open at high speed by a cam placed on the track. When Stapp finally made the run, the doors opened on cue, exposing Stapp to a blast of air estimated to be moving at more than 200 m/second. By all accounts, it was Stapp's easiest sled run, which may have prompted him to make one final ride to really push the envelope: a transonic run at Mach 0.9. This time, Stapp would be protected only by a helmet and visor and, when the sled stopped, Stapp would establish the threshold of human survivability to G-force – or be killed in the process. It was this latter possibility that concerned his friends and co-workers who couldn't understand the justification for risking being killed. To document the proceedings, Stapp asked Joe Kittinger, who was involved in some of Stapp's other experiments (flying zero-G profiles to study the effects of weightlessness), to fly a T-33 photo chase plane for the run. Kittinger describes his meeting with Stapp in his autobiography *Come Up and Get Me* [1]. "Sit down, Captain," Stapp said when Kittinger showed up in his office. "I've got a deceleration study going on I want to talk to you about. Next month we're going to fire off nine rockets attached to an aluminium sled and send it down a 3,500 foot track. At the end of the track, we'll go from top speed in the vicinity of Mach 1 to a full stop in about one second." Even Kittinger, accustomed as he was to Stapp's madcap ideas, found it hard to believe what he was hearing. "I need photographic documentation – aerial photos – and I want you to fly the photographer. Are you interested?" Kittinger, always up for a challenge, accepted immediately, not knowing at the time that the run would be manned. On December 10th, 1954, Stapp was fitted with a rubber bite block and equipped with an accelerometer. Just seconds before ignition, a siren shrilled and two red flares lofted skywards. Overhead, Kittinger approached in his T-33, pushing his throttle wide open in preparation for the launch. At launch, the Sonic Wind's nine rockets detonated, spewing 10-m rooster tails of fire, sending Stapp rocketing down the track. "He was going like a bullet," Kittinger remembers. "He went by me like I was standing still, and I was going 350 mph." Just seconds into the run, the sled reached its peak velocity of 1,011 km/hr (actually faster than a bullet), subjecting Stapp to 20 G. Kittinger watched dumbstruck, thinking there was no way the sled could stop in time. But then, just as it had been designed to do, the sled hit the water brake and the rear of the sled, its rockets expended, tore away, allowing the front section to continue downrange before slowing when it hit the second water brake. Spray formed a maelstrom from the back of the Sonic Wind. From Kittinger's vantage point, it appeared as if the sled had hit a concrete wall. "He stopped in a fraction of a second. It was absolutely inconceivable that anybody could go that fast and then just stop, and survive." Kittinger's thoughts were shared by the ground crew, who rushed to the scene, half expecting to find a corpse in the sled. Instead, they were relieved to find Stapp smiling, despite being in obvious pain and, with eyes that had filled with blood, looking like an extra in a horror movie:

> "Following the run I witnessed, all the blood vessels in his eyes had burst, and he was a frightful sight for a few days, but in typical fashion he never stopped laughing about it. All the rest of us could do was shake our heads in awe."
>
> Joe Kittinger, describing Stapp's 1,000-km/hr run in his autobiography, *Come Up and Get Me* [1]

When the Sonic Wind had hit the water brake, it had produced 46.2 G of force and, for just over a second, Stapp had been subject to 25 G, which was equivalent to a Mach 1.6 ejection at 13,000 m, or crashing your car into a brick wall at 200 km/hr. Not surprisingly, the force had burst nearly every capillary in Stapp's eyeballs. Even Stapp was concerned this time, worrying that perhaps he had finally pushed his luck too far, muttering about having to spend the rest of his life walking with a white cane. Fortunately, his vision had returned more or less to normal by the following day. What didn't return to normal was his life. His ride on the Sonic Wind catapulted him into the mainstream of celebrity. He was dubbed "The Fastest Man on Earth", graced the pages of *Collier's* and *Life* magazines, he was featured on the cover of *Time* magazine in 1955, and was even the subject of a Hollywood B-movie. Stapp used his newly found fame to good effect, urging car manufacturers to look at his crash data and design cars with safety in mind. He also lobbied hard for the installation of seat belts and improvements such as soft dashboards, collapsing steering wheels, and shock-absorbing bumpers. Later, as the era of manned spaceflight approached, Stapp told the American Rocket Society that experiments using the rocket-powered sled could accelerate the goal of flying astronauts into space and he subsequently helped conduct tests on human and animal subjects in the Johnsville Centrifuge (Chapter 4).

While he was promoting car safety, Stapp announced plans to make a Mach 1 sled run but his superiors didn't think it was a good idea. For once, they were right. In June 1956, during an 80-G test, the Sonic Wind left the track and was severely damaged. Human tests were immediately suspended and Stapp's days as a rocket man were over. Not that the great man was too disappointed. After all, he had already proven that a pilot, if properly protected, could survive a high-speed, high-altitude ejection. He had also defined, or at least come damned close (after all, any closer and he probably would have been killed) to defining, the human limit for deceleration. But, while the experiments had found what could be called a tolerance limit for windblast, it hadn't found the lethal point, which was why animal experiments continued, mostly using chimps. During the tests, chimps suffered varying degrees of injury, mostly minor, depending on the type of harness and protective covering worn, but there was no indication that even the highest level of windblast experienced was necessarily injurious to a properly secured and protected subject. So, the logical next step was to have the sled travel at even greater velocities. At the time, an extension of the Holloman track to 1,500 m was under consideration, which would have helped the planned tests. But the extension hadn't been finished when another construction project was started, designed to extend the facility to 10,600 m, which would make it the longest in the world. The 10,600-m track wouldn't be ready for many months though, so Stapp and his associates transferred the windblast test operations (and Sonic Wind No. 2) to the Supersonic Naval Ordnance Research Track at China Lake.

Assisting Stapp at China Lake was Doctor John D. Mosely, Chief of the Biodynamics Branch. Mosely's first windblast test, on February 18th, 1957, was the first at China Lake and also the first high-speed track experiment since March 2nd, 1956, that was designed to assess windblast. It was a checkout run that reached a velocity of 403 m/second. The first full-scale experiment took place on April 13th, with very moderate acceleration and deceleration and a peak velocity of 590 m/second. During the run, the test chimp wore a special flying suit devised by the Aeromedical Field Laboratory and a helmet. Unfortunately, the headrest failed even before the sled reached supersonic speed, the helmet failed in turn, and the chimp's head was yanked so violently that it broke the unfortunate animal's neck.

The next run at China Lake was held on June 27th and reached 577 m/second, with a duration of two seconds at roughly Mach 1.7. Maximum windblast was about 167,581 pascals. Once again, the test resulted in the subject's death but, this time, it occurred 24 hr after the run and the cause was different. The chimpanzee was adequately secured against flailing, but the flying suit tore, exposing the subject to serious burning from windblast. At least the chimp fared better than three guinea pigs attached to the same test sled by the Bio-Acoustics Branch of Wright Air Development Center's AML. These unfortunate test subjects were attached with nylon netting, and the third was placed in a metal container whose largest opening measured one inch by two inches. The can itself stood up through the test, but all three guinea pigs vanished into thin air.

Despite guinea pigs vanishing into thin air and the high death toll of the chimps, Stapp and Mosely were confident that the burning could be avoided by using Dacron sail cloth. When the next test took place on March 12th, 1958, they were proved right. Although the subject died due to harness failure, the new material proved satisfactory. Stapp and Mosely conducted three more tests to make sure. Although the wind pressure in the last three tests didn't reach the highest levels expected in an operational escape situation, the levels reached were encouraging when the altitude of flight operations was taken into account (at higher altitudes, the air density, and thus potential wind pressure for any given speed, is lower). Ultimately, the tests demonstrated that real progress had been made in devising a means of protection for an ejecting crewman.

With human rocket tests complete, Stapp handed the work over to the engineers to ensure humans were kept inside the envelope that he had worked so hard on and risked so much to define. And that's precisely what they did. Thanks to Stapp's data, engineers were able to develop new generations of aircraft capable of flying higher, faster, and safer than any ever built. They also built safer cars thanks to Stapp convincing the Air Force to build a car crash test facility, where the first-ever car crash tests (with dummies) were conducted. And, in 1966, when President Lyndon Johnson signed a law requiring seat belts in all new cars, Stapp was by his side. By putting his life on the line, Stapp had turned current engineering safety standards around and had demonstrated that, if appropriately positioned and secured, the human body could endure amazing crash forces. Stapp went on to champion his knowledge and research into the automotive domain, bringing about many safety changes and establishing a conference that still bears his name. Ultimately, John Stapp put a face on exploring this limit of human physiology and opened up a world of research in the unusual branch of biodynamics.

Reference

[1] Kittinger, J.; Ryan, C. *Come Up and Get Me: An Autobiography of Colonel Joe Kittinger*, pp. 42, 44–45. University of New Mexico Press, Albuquerque (2010).

2 To Black Out or Not to Black Out

"At nine G's, you can't move a muscle. It pins your hands and arms down, and your head weighs 100-plus pounds."

Major Matt Modleski, an F-16 pilot (174th Fighter Wing)
in Syracuse, NY, and former lead solo pilot for the
Thunderbirds aerial demonstration squadron

Acceleration (G) is one of the most demanding aspects of flying a high-performance fighter jet. As many as one in 10 pilots reports episodes of gravity-induced loss of consciousness (G-LOC – pronounced "gee-lock") while flying aerobatic maneuvers, although the real incidence is probably higher because G-LOC often causes amnesia. The unconsciousness that occurs during G-LOC is similar to fainting but, for the pilot, the consequences can be lethal; the US Air Force (USAF) reported G-LOC as the cause of 18 fatal accidents between 1982 and 1990.

To better understand why G-LOC poses such a threat to pilot safety, it's helpful to understand what happens to the body when it's subjected to positive G (abbreviated +G). The most significant effect of +G on the pilot's brain and eyes is a reduction in blood pressure and blood flow. The eye reacts first. As G-forces increase and blood pressure drops, aircrew experience grayout (loss of color and clarity), tunneling of vision, and blackout, although blackout shouldn't be confused with loss of consciousness because a pilot can be blacked out and still be flying the aircraft, albeit not very well! However, if Gs continue to increase beyond blackout, the pilot will G-LOC because when the brain loses its blood supply and exceeds its oxygen reserve, it abruptly fails. Once it fails, it stays "turned off" for a variable length of time, even after blood flow is restored. Consciousness can be maintained when the G-onset rate is slow (say 1 G/second) – enough for visual symptoms to be recognized – but, when G-onset rates are high (5 or 6 G/second) and the peak G-level sustained is high, as is often the case in modern fighters (Figure 2.1), G-LOC can occur without any visual warning signs. G-LOC is a serious problem because no aircrew is immune. It can happen to anyone flying any aircraft at any time when under G-stress.

Once a pilot G-LOCs, their brain enters a state known as *absolute incapacitation*, which lasts for about 15 seconds (the range is anything from 5 to 30 seconds). This state will occur even if the pilot has unloaded the G. While incapacitated, the pilot is in a dream-like state, unaware of his or her environment, unable to respond to any outside stimuli and, most importantly,

Figure 2.1 The F-35 Lightning II is a single-seat, single-engine, fifth-generation multirole fighter under development capable of generating 6-G/second onset rates. Image courtesy: USAF.

not in control of the aircraft! The brain's blood supply returns during this period and gradually the brain starts to "wake up" and enters a period of *relative incapacitation*, which lasts about another 12–15 seconds. The combined absolute and relative incapacitation times are referred to as the *total incapacitation time*, a period that averages about 27 seconds. At the end of this total incapacitation period, the pilot is able to recognize where they are and respond to the environment. Occasionally, aircraft recovery may be possible during the relative incapacitation period, as the pilot may respond to directive calls such as "pull-up, pull-up".

The third phase of recovery from G-LOC is the recovery of cognitive processing skills, which may require several minutes before a return to full function (Figure 2.2) is realized. During this period, functional motor skills and situational awareness may be severely impaired, so much so that an attempt to suddenly recover the aircraft may induce a subsequent G-LOC episode. By now, the implications of G-LOC and its after-effects in a combat scenario are obvious. The pilot may be conscious after a G-LOC, but they won't be in any state to fly the aircraft.

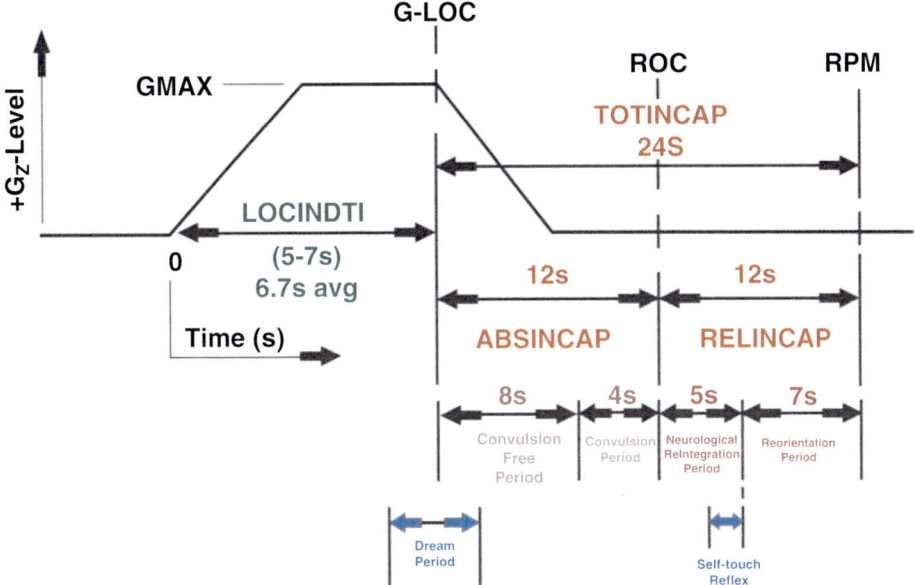

Figure 2.2 The time relationship of G-LOC syndrome events. LOCINDTI = loss of consciousness induction time; G-LOC = onset +Gz-induced loss of consciousness; ROC = return of consciousness; RPM = return of purposeful movement; ABSINCAP = absolute incapacitation period; RELINCAP = relative incapacitation period; TOTINCAP = total incapacitation; convulsion-free period = period during unconsciousness where myoclonic convulsions do not occur; convulsion period = myoclonic convulsion period; reintegration period = period where neurological reintegration occurs following loss of consciousness; reorientation period = period where reorientation to the environment occurs ending in the return of purposeful movement; dream period = period where dreamlets often occur. The length of unconsciousness and the associated incapacitation are usually dependent on the magnitude of the central nervous system insult resulting from reduced blood flow. The magnitude of the insult is usually determined by the onset and offset rates of the +Gz stress and the length of time at increased +Gz. Incapacitation resulting from +Gz stress include between 10 and 15 seconds of absolute incapacitation and 10–15 seconds of relative incapacitation (confusion/disorientation) for a total of 20–30 seconds of total incapacitation. Image courtesy: James E. Whinnery, Ph.D., M.D. Aeromedical Research Division, Civil Aerospace Medical Institute.

How does the pilot feel when they recover from G-LOC? First of all, the pilot may not even recognize the G-LOC incident because one of the symptoms of G-LOC is partial amnesia caused by impaired oxygen flow to the brain. However, once the pilot checks their instruments and heads-up display (HUD) and notices the inexplicable sudden loss of altitude, they will acknowledge that they've suffered a G-LOC incident. As they gradually recover from the discombobulating feeling of regaining full consciousness, they may also experience tingling around the mouth or in the extremities and perhaps a sense of dreaming, although some pilots report nightmares as a symptom. In fact, pilots suffering from G-LOC experience a constellation of symptoms which are defined collectively as the G-LOC syndrome:

1. Loss of peripheral vision
2. Tunnel vision
3. Blackout (complete loss of vision)
4. Loss of consciousness
5. Loss of motor control (purposeful movement)
6. Loss of sensory input to the brain
7. Lack of memory formation
8. Myoclonic convulsions
9. Dreamlets
10. Recovery of consciousness
11. Neurological reintegration
12. Neurological external environment reorientation
13. Return of purposeful movement
14. Transient tingling or slight numbness of the extremities
15. Alteration of psychological state (anxiety, confusion, embarrassment)

While a G-LOC event may cause a pilot to crash, the syndrome is actually a safety mechanism that has evolved to protect us in a gravitational field and to ensure the optimum protection of the organ system that is the key to its evolutionary success on Earth: the brain. And it's not as if the brain gives up at the slightest hint of G-stress. Far from it. As soon as G-stress is detected, the cardiovascular and neurological systems initiate protective reflex mechanisms so functional compromise does not occur easily. In fact, a significant increase in +Gz stress above +1 Gz must be applied before any of the G-LOC symptoms occur. But, when the Gs become excessive, visual symptoms of grayout, tunnel vision, and blackout warn the pilot that the cardiovascular and neurological reflex responses are inadequate for the magnitude of the stress and that evasive action is required immediately. If evasive action is not taken, G-LOC occurs, but it only occurs when the brain becomes threatened by a lack of oxygen (hypoxia) and can't function reliably.

   With the pilot unconscious, the brain is in a minimal-energy-expenditure state, with loss of sensory, motor, and consciousness function. If a G-LOCed pilot were to be hooked up to an electroencephalogram (EEG), a flight surgeon would observe a synchronized slow-wave pattern that would persist until the recovery process started – the relative incapacitation period discussed earlier. At this stage, blood flow would begin to return to normal levels and myoclonic twitching might occur. You may wonder why a pilot would start convulsing but the mechanism, like all the G-LOC recovery processes, has a purpose. The twitching serves to contract the muscles in the extremities and abdomen, thereby enhancing return of blood to the central circulation and ultimately the brain. And the dreamlets and nightmares? Well, scientists think the dreamlets serve to alert the pilot that G-LOC has occurred. Without the dreamlets, pilots may not recognize G-LOC episodes and, if such events go unnoticed, the pilot may not recognize the importance of future threat avoidance when another G-LOC event occurs. Ultimately, the relative incapacitation period serves to ensure all the sensory, motor, and consciousness functions are reintegrated and ready to be utilized, which is a good thing because the pilot will most likely need all their senses just to survive when they fully recover!

G-LOC

G-LOC was first reported way back in 1918 during air races, when it was known as "fainting in the air". The problem continued to be reported by pilots after the First World War, as aircraft designs improved and concerns related to G-LOC led to operational restrictions on the turn rates of some aircraft during the 1920s. The problem became even more acute in the 1930s when aggressive dive-bombing techniques were developed that resulted in severe acceleration during the pull-up at the end of the bombing run. The problems were correctly diagnosed by Royal Air Force (RAF) flight surgeons as "cerebral anemia produced by centrifugal action". Based on their observations, the flight surgeons reckoned that 4 G was the limit of human acceleration tolerance and methods were investigated to help the pilots deal with the Gs encountered during the dive-bombing runs. One of these methods was a pneumatic acceleration belt developed by the US Navy in 1934. The belt was inflated by the pilot prior to the dive-bombing run, but the device only had a marginal effect on acceleration tolerance, so research continued. The research was made possible thanks to the development of human centrifuges designed specifically for aviation-related studies. Thanks to extensive centrifuge research during the 1940s by Dr Wilbur Franks, the first workable anti-gravity suit (the suit was developed under the name "Franks Flying Suit" by Franks and his colleagues at the Banting and Best Medical Research Institute at the University of Toronto) was developed in 1941. Made with rubber and water-filled pads, the suit (Figure 2.3) counteracted the effects of high G-forces on aircraft pilots and was used during the Second World War (G-suits worn by air force pilots as well as astronauts and cosmonauts are based on Franks's original designs).

Later developments of the G-suit resulted in the five-bladder pneumatic version (also called "G trousers") but the problem of G-LOC remains today due to the increased engine thrust and wing-load capacity of current aircraft that allow sustained accelerations of 9 G or more. In fact, G-LOC has become such a problem that, for the first time since the 1920s, the pilot's limitations began to restrict aircraft maneuverability – a serious problem in aerial combat situations where the ability to turn hard and climb fast may be the key to survival. Furthermore, sudden onset of high acceleration can reduce the pilot to

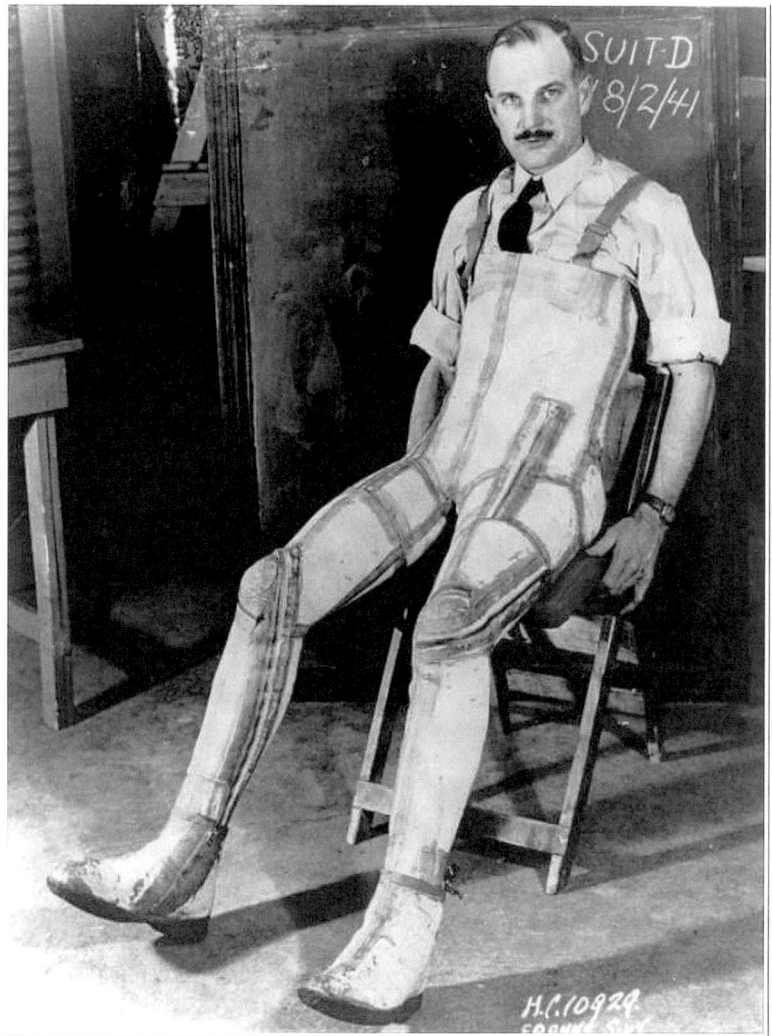

Figure 2.3 Franks Flying Suit. Wilbur Rounding Franks was born on March 4th, 1901, in Weston, Ontario. He attended the University of Toronto, graduated with a degree in medicine, and trained in cancer research. While Sir Frederick Banting was the first person to recognize the G-force complication, it was Franks, in 1941, who designed and tested the first-ever G-suit – the Franks Flying Suit Mark II. The suit comprised two layers of rubber overalls laced together, with liquid between each layer. It was capable of protecting pilots from a maximum of 6 G. Franks later designed a suit that functioned by air compression, instead of liquid compression. This suit was successful and went into production, and is still in use today. During the Second World War, Franks also fabricated the first Canadian human centrifuge and used the system to train pilots to operate their aircraft while encountering high G-forces. Image courtesy: DRDC.

unconsciousness without warning. In response to these problems, research/ training centrifuges have been upgraded to achieve the rapid onsets (6 G/second) and high accelerations (12 G) that new aircraft are capable of. Recent developments have included extended-coverage anti-G suits and systems for balanced pressure breathing to better counteract the circulatory effects of high acceleration. Also, several nations now use centrifuge training to better train their pilots in the G environment. But, despite these efforts, G-LOC continues to present a problem because there are so many factors that can affect the pilot when they are subjected to changing acceleration. To appreciate just how much G exerts its effect on pilots and how this influence can be mitigated, it's necessary to have a basic understanding of G physiology. But, before we do this, it's useful to understand some of the conventions and terminology (Table 2.1)

Table 2.1. Terms used in the high-acceleration environment.

Term	Definition
g	Acceleration equal to gravity at the surface of Earth, 9.80665 m/s^2
G	A unit of convenience calculated as the observed acceleration divided by g; thus, acceleration of 29.4 m/s^2 is expressed as 3 G
Grayout	Dimming of vision due to reduced retinal perfusion; usually accompanied by narrowing of visual fields (tunneling)
PLL	Peripheral Light Loss: loss of vision at the edge of the field of vision
Blackout	Loss of vision during acceleration due to insufficient retinal perfusion; precedes loss of consciousness because eye perfusion is opposed by normal intraocular pressure
Redout	Reddish fogging of vision due to venous pooling and increased perfusion pressure in the eye during exposure to –Gz
G-LOC	Acceleration-induced loss of consciousness during sustained acceleration due to inadequate cerebral perfusion
Anti-G suit	Trousers fitted with pneumatic/hydrostatic bladders over the abdomen and legs; an inelastic outer layer ensures increased bladder pressure during +Gz is transmitted to the adjacent tissues to minimize venous pooling
Pressure breathing	Continuous positive pressure applied to the airway during +Gz to increase intrathoracic pressure and thereby raise arterial blood pressure
Balanced pressure breathing	Use of a bladder contained in a vest to reduce the work of pressure breathing. The bladder is inflated to airway pressure and prevents over-expansion of the chest as well as making it easier to exhale against pressure; also called "assisted" pressure breathing
Anti-G straining maneuver	Voluntary isometric contraction of major muscle groups to prevent venous pooling and preserve cerebral perfusion; valsalva maneuver used to increase intrathoracic pressure and arterial pressure to preserve cerebral perfusion

used when discussing acceleration and its physiological effects. It's also helpful to remind ourselves of the acceleration vectors that were discussed in Chapter 1. Remember, the dominant acceleration for a seated pilot is +Gz (head-to-foot), which occurs in ordinary turns and pull-ups, and forces blood to pool in the legs and feet. Its reverse, –Gz (foot-to-head), accompanies outside loops and forces blood towards the head. Pilots can also be subject to Gx (front-to-back) and Gy (side-to-side).

G physiology

With G-forces acting in so many directions, it's not surprising the body gets a workout. For example, +Gz exerts mechanical effects on soft tissues and compresses the spine, and it also affects the cardiovascular and pulmonary systems, creating risks of visual symptoms and G-LOC. –Gz is just as bad, as it causes visual and cardiovascular disturbances and is just as capable of causing G-LOC. Then there are the mechanical effects of acceleration, which causes soft tissues to sag, with the result that a person subject to G appears to have aged prematurely. Fortunately, it's a reversible change. The bottom line is that the sheer magnitude of G in any axis causes problems. Above 2.5 G, most people find it difficult to rise from a seat and, when that acceleration increases to 3 G or more, raising an arm is a workout. Crank up the Gs some more and, at +8 G, any gross movement is next to impossible, even if your name is Arnold Schwarzenegger! Being subject to such high G-levels causes serious problems for pilots because their helmet alters the center of gravity (CG) of the head and their acceleration-magnified weight puts tremendous stress on their neck muscles. It also temporarily compresses the spinal column by as much as 5 mm. A helmeted pilot pulling +8 Gz can keep their head erect but, if it tips forward, the chin drops onto the chest and can't be raised until acceleration is released, which is not a good position to be in when flying a high-performance jet.

In tandem with the mechanical effects are the hydrostatic effects, since acceleration increases the 'weight' of the blood, thereby increasing the pressure gradient in the hydrostatic column. This creates havoc in the cardiovascular system. A simple example is a standard turn in which the pilot is oriented in the +Gz direction, which means the pressure gradient is increased from head to foot. This increase in the pressure gradient directly increases venous pressure and pooling below the heart as the turn forces blood towards the feet. This force results in less blood flowing to the heart and regions above the heart – most importantly, the brain. In fact, it is the heart-to-head distance that has a significant negative correlation with acceleration tolerance, which means tall individuals have a slight disadvantage when it comes to pulling G. At high Gs, in the +9-Gz region and above, the pressure causes effects such as *petechiae* (Figure 2.4), often referred to as "G-measles" – one of several occupational hazards among fighter pilots.

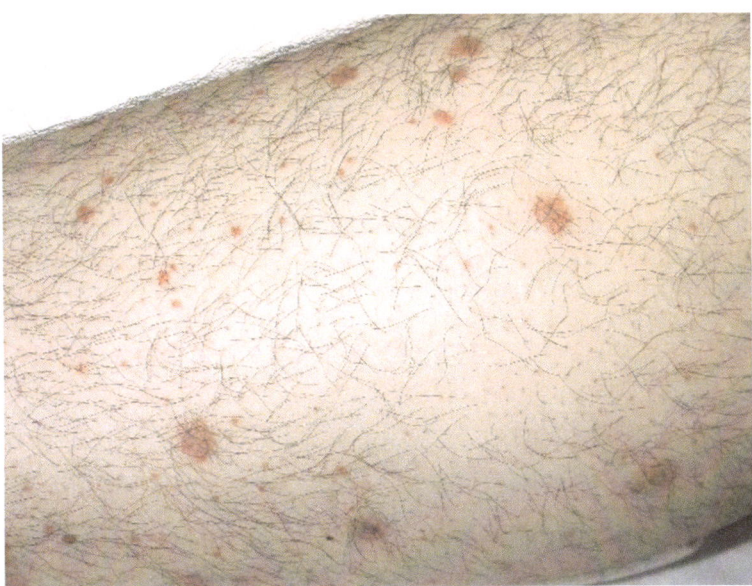

Figure 2.4 One of the stresses to which pilots are subjected in highly maneuverable military aircraft is produced by high-speed turns. The resultant elevated G-forces may cause petechial hemorrhages, particularly in the lower limbs. The phenomenon usually begins to develop in healthy individuals at approximately +5 Gz, with virtually everyone having petechial hemorrhages at +9 Gz. Individuals develop an acclimatized resistance to these formations, whose exact etiology is unknown. Fit problems with helmets, anti-G suits, and vests which seem minor on the ground may lead to painful pinching, friction, and bruising at high acceleration. The increased intravascular pressures in the body during high G-loads, straining, and pressure breathing may also cause petechiae. Image courtesy: www.primehealthchannel.com.

While petechiae is little more than an unsightly problem, the lack of blood going to the brain and the eyes may be deadly. A pilot who loses vision is a pilot in trouble. While the body has some compensating mechanisms to help overcome the effects of acceleration (e.g. pressure receptors located above the heart detect low pressure and compensate with an increased heart rate), the eyes have no such protection, which is why the onset of acceleration typically produces visual symptoms before G-LOC. +Gz also makes breathing difficult by pulling down the diaphragm and collapsing the air sacs in the lungs (the greater the G, the more air sacs collapse, like a balloon collapsing), causing a G-induced symptom known as *acceleration atelectasis*.

–Gz isn't much kinder. For example, when a pilot performs an outside loop or transitions from a steep climb into a dive, the maneuver forces blood towards the head, with the effect that the pilot feels like their head is about to explode. Along with the feeling of pressure inside the head, pilots subjecting themselves to high negative Gs suffer facial edema, blurred vision, and a red-colored visual fog –

redout. Not surprisingly, a high level of –Gz is extremely uncomfortable and pilots will tell you that their eyes feel as if they're about to pop out. Along with the "eye-popping out of your head" sensation, negative G also subjects pilots' cardiovascular systems to various cardiac arrhythmias such as bradycardia (slowing of the heart rate), which may ultimately lead to unconsciousness. While a stress of –2 Gz may be tolerated for up to five minutes, –3 Gz can be borne for only 30 seconds and –4 Gz for only a few seconds. Any longer and pilots inevitably G-LOC. But, like so many extreme environments, there are those who perform better than others, which brings us to the subject of tolerance limits.

Acceleration tolerance

There are myriad factors that combine to determine G-tolerance on a given day. For example, short pilots have a higher tolerance than tall pilots due to the respective differences in heart-to-brain distances. Individuals with higher blood pressure have a higher tolerance than those with lower blood pressure. Being dehydrated can also reduce G-tolerance. Acclimatization training is also an important factor in G-tolerance. For example, most military pilots are required to undergo centrifuge high-G training to ensure they are proficient in performing protective anti-G straining maneuvers (AGSMs), which we'll discuss shortly. Finally, a good understanding of factors relating to G-tolerance, especially the G-LOC syndrome, is vital for all who enter the high-G environment. Having said that, there are certain factors that may limit an individual's ability to tolerate G.

An abnormality of the neurological or neurovascular system will most likely preclude an individual from engaging in high-G activity due to the potential for sudden incapacitation, since any such abnormality might contribute to compromising blood supply to the brain. Since the cardiovascular system is the system primarily affected by +Gz, any abnormality in cardiovascular anatomy or physiology is reason for concern in aerospace safety. Equally, medications that alter cardiovascular physiology are viewed with caution, especially pharmacological agents that alter blood pressure and the function of the heart. The heart is particularly susceptible because acceleration is a *dysrhythmogenic* stress, which means that anything that affects cardiac rate or rhythm is a threat to safety. Flight surgeons have a name for the group of symptoms that may affect the heart during +Gz: *tachydysrhythmias*. Like most medical terms, tachydysrhythmia appears to have been borrowed from another language but, in layman terms, it simply means a quickening of heart rate (ventricular tachycardia) and premature beats (supraventricular and atrial) which are common during +Gz. For the symptoms that occur following +Gz, flight surgeons have another tongue twister: *bradydysrhythmias*. Like its counterpart, this term has a simple explanation, describing the out-of-sequence beats (sinus arrhythmia), bradycardia, and spontaneous heart beats (ectopic atrial rhythm) that occur following +Gz.

Musculoskeletal problems, particularly those affecting the neck and back, are of particular concern during +Gz stress. For example, any anatomical abnormality that decreases neck or spinal strength or stability has to be carefully considered. It's the reason pilots are prescribed neck and back muscle-strengthening exercises to prepare themselves for the G environment. The pulmonary system is also significantly affected by +Gz stress; blood can be drawn away from the lungs, resulting in less oxygen being delivered to the muscles and to the nervous system.

Tolerance limits for +Gz acceleration are usually signaled by visual symptoms such as peripheral light loss (PLL), tunneling, grayout, and blackout. During gradual onset of acceleration, a relaxed subject not wearing an anti-G suit typically experiences initial visual symptoms at about +4 Gz, although susceptible individuals may experience PLL as low as +2 Gz. "G monsters", on the other hand, may not notice anything until +7 Gz. In my job as director of Canada's manned centrifuge operations, I've witnessed about half a dozen such individuals, one of whom casually chatted away as the G-level crept over the 7-G mark! For most people, once they've experienced initial visual symptoms, the next symptom will likely be blackout at about +5 Gz and unconsciousness if the Gz continues to increase and no strain is attempted. Once the individual has lost consciousness, they are deemed to have G-LOCed. Witnessing an episode of G-LOC can be a little disconcerting. First, the hapless individual's head drops to the chest and seizure-like flailing motions may occur – some people have broken limbs as a result of this. While their limbs are flailing, the now-unconscious individual may exhibit myoclonic, spastic-like twitching. Fortunately, consciousness returns quickly (a typical G-LOC period is 15 seconds) and the individual will slowly raise their head, looking very, *very* confused. This confusion is quickly exacerbated when they realize they don't remember the incident and, when asked if they know what happened, some even deny losing consciousness. If the G-LOC incident occurred in a centrifuge, there is a video record that the individual can watch afterwards to prove otherwise! Repeated episodes of G-LOC in healthy individuals doesn't appear to have any serious side effects, but it's not something a pilot wants to experience while flying a high-performance fighter jet! During flight, a G-LOC episode means the pilot will unexpectedly cease to control the aircraft for a critical period of time and may not realize what has happened until it's too late.

While G-LOC is a cause of a small percentage of fighter jet accidents, its role in a fatal accident is difficult to determine because G-LOC is a transient functional state, so there is no evidence available after the event. What *is* known is that G-LOC accidents are characterized by the crash of a mechanically sound aircraft shortly after performing high-G maneuvers with a lack of appropriate pilot response for recovery from the situation. This lack of response is often evidenced by the abrupt cessation of voluntary straining maneuvers in cockpit sound recordings.

Crash investigators who suspect G-LOC as the reason for a crash often zero in on factors such as fatigue, sleep deprivation, hangovers, illnesses, or anything that might conceivably reduce acceleration tolerance. Other factors that have been implicated in G-LOC incidents include heat stress and dehydration, which measurably decrease acceleration tolerance, owing to the combined effects of increased peripheral vasodilation and reduced plasma volume. Also linked to G-LOC are factors such as hyperventilation due to anxiety, mental stress, hypoxia, and pressure breathing, which may decrease acceleration tolerance through cerebral vasoconstriction and peripheral vasodilation.

Countermeasures to G

With so many factors influencing an individual's tolerance to acceleration, it's not surprising that scientists have dedicated a lot of time to investigating how to combat G effects. One of the ideas they came up with was using semi-reclining seats (obviously to the extent that they were compatible with the aircraft design).

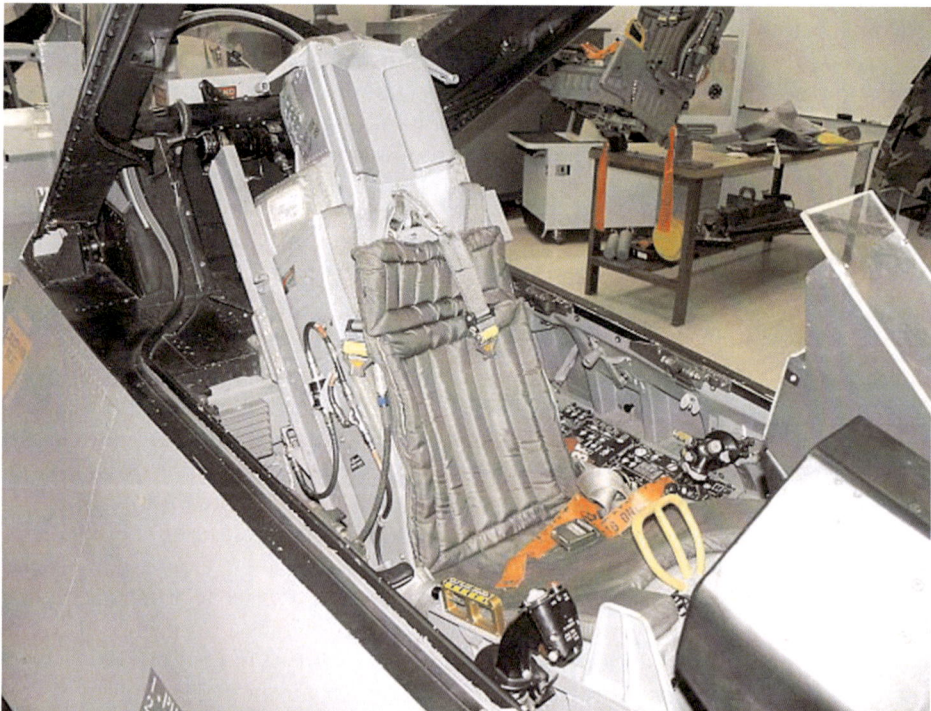

Figure 2.5 The F-16's Advanced Concept Ejection Seat (ACES) II is reclined at a tilt-back angle of 30°. Most fighter aircraft have a tilted seat at 13–15°. The seat angle was chosen to improve pilot tolerance at high G-forces and reduce susceptibility to gravity-induced loss of consciousness. Image courtesy: USAF.

Now, you may be wondering how changing the angle of the pilot's seat could influence acceleration tolerance, but the idea is rooted in acceleration physiology. You see, the standard cockpit seat has the whole length of the pilot's hydrostatic column subjected to +Gz acceleration, which doesn't help the pilot tolerate acceleration at all. But, alter the seat position to a 30° seat-back angle as the USAF did with their F-16 (and the French with their Rafale fighter) planes (Figure 2.5) and you reduce the heart–brain distance, which leads to a small but measurable increase in the threshold for visual symptoms. Of course, greater effects would be achieved by further reclining the seat to 70° and elevating the feet or placing the pilot in the prone position, but imagine trying to fly a high-performance jet in that position!

While adjusting the seat angle provides a small increase in acceleration tolerance, greater increases can be achieved through skilled use of the AGSM, which requires the pilot to simultaneously contract the muscle groups of the abdomen and legs, while inhaling sharply and holding their breath for three seconds at a time.

Anti-G straining maneuver

The AGSM is *the* best G-defense measure available to aircrew members. While there are all sorts of equipment countermeasures such as the anti-G suit and COMBAT EDGE, which we'll discuss shortly, this equipment was never meant to replace the AGSM, only aid it. Since the AGSM is so important to aircrews, it's worth taking a look at its components. To begin with, it's useful to know just how effective a good AGSM can be by considering that the average aircrew's relaxed G-tolerance in an F-16 seat is about 5.2 G. Now, the G-suit, which will be discussed shortly, can add another 1.6 G of protection, and a good AGSM can add another 3.5 G or more of tolerance. When these are totaled, one can appreciate that 9G, which is the performance envelope of many fighters (Figure 2.6), is a big challenge for most aircrew, with little or no safety margin.

The first step pilots learn is that the straining maneuver should precede rapid G onset. As so many pilots learn from painful experience in the centrifuge, it's extremely difficult to "catch up" to a G-load if they get behind from the start. When I'm instructing pilots in the performance of the AGSM, I always advise them to start with a 100% effort and dial down the effort if they can. Another important component is recency. If a pilot has been 'flying' a desk for several months, they will most likely have suffered some AGSM deconditioning and won't be prepared for the high +Gz environment.

The AGSM can be broken down into three components: maximal contraction of lower body muscles, forced exhalation against a closed glottis, and a three-to-one count. Learning the three components can be a bit of a juggling exercise for some pilots, some of whom focus too much on one or two of the components, quickly realizing that, without all three elements, the strain's effectiveness is significantly reduced. The AGSM may seem like a strange combination of drills

Figure 2.6 The Mikoyan MiG-29 is a fourth-generation jet fighter aircraft designed in the Soviet Union for an air superiority role. Image courtesy: Wikimedia Commons.

but each component serves a purpose in increasing a pilot's G-tolerance. The muscle tensing increases usable blood volume and return of blood to the heart; the tighter the muscles are contracted, the greater the reduction in blood pooling, the greater the blood pressure, and the greater the return to the heart. The key to straining the muscles is maintaining it continuously, even when breathing. If the muscles are relaxed while under G, the blood will immediately rush into the extremities and the pilot will find it almost impossible to catch up, which may result in the dreaded G-LOC (Figure 2.7).

The increased chest pressure increases heart output pressure and works as a sort of boost pump for the heart. The greater the pressure generated in the chest, the more the heart and blood vessels leading from it are squeezed, which keeps brain blood pressure where it needs to be. Perhaps the component that pilots have the most trouble with is the breathing cycle. At the onset of G, pilots are instructed to take a deep breath, close the glottis (throat), and bear down with the chest muscles as if they are trying to exhale, but keeping the throat closed. The cycle of breathing generates intrathoracic pressure, which is important in sustaining blood pressure. The problem pilots often make is to breathe too quickly or too slowly. The optimal rate is taking a breath every 2.5–3.0 seconds, then exhaling a small amount of the air, and immediately pulling the air back in to regenerate the chest pressure. The exhalation and inhalation process ideally takes about 0.5 seconds. If the exhalation/inhalation cycle is too slow or too fast, the blood pressure starts to fall and the effectiveness of the AGSM starts to degrade. When we train pilots in the centrifuge, we videotape their runs so we

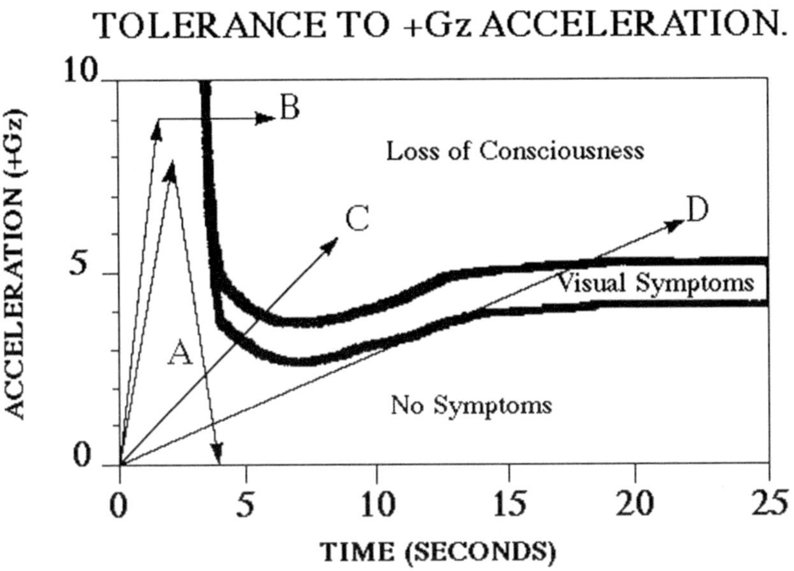

TOLERANCE TO +Gz ACCELERATION.

Figure 2.7 G-LOC, pronounced "gee-lock", is the abbreviation of G-induced loss of consciousness, a term that describes a loss of consciousness occurring from excessive and sustained G-forces draining blood away from the brain. Image courtesy: FAA/NASA.

can coach them on their mistakes following the run. The most common coaching point is timing, while the second most common problem is failing to anticipate the G. In the centrifuge, the pilots are counted down to the start of the run so they have plenty of warning but, in flight, when performing combat maneuvers, there is less time to anticipate the G. This is important because, to maximize the effectiveness of the AGSM, it should begin before G is loaded on the aircraft – failure to do so results in the aircrew member either trying to catch up on the AGSM (a very dangerous practice) or having to unload in order to buy time to catch up. Perhaps the third most common mistake is not grading the strain. For example, performing a strain with the intensity necessary to stay awake at 9 G when the G-load is only 5 G will result in early fatigue and increase the likelihood of G-LOC in subsequent engagements, so pilots are coached to grade the intensity of the AGSM in relation to the level of G. This is why it's always safe to overestimate the intensity of the strain and always unsafe to underestimate the intensity.

While the aircraft and personal G-protection equipment are passive, the AGSM is active. It requires anticipation of the maneuver and is a practiced skill that must be integrated with the myriad cockpit tasks. Like all athletic skills, the efficiency and intensity of the AGSM depend on multiple factors, including strength, endurance, training, motivation, and proficiency. Like other athletic skills, it is also susceptible to deconditioning – studies have shown there is some

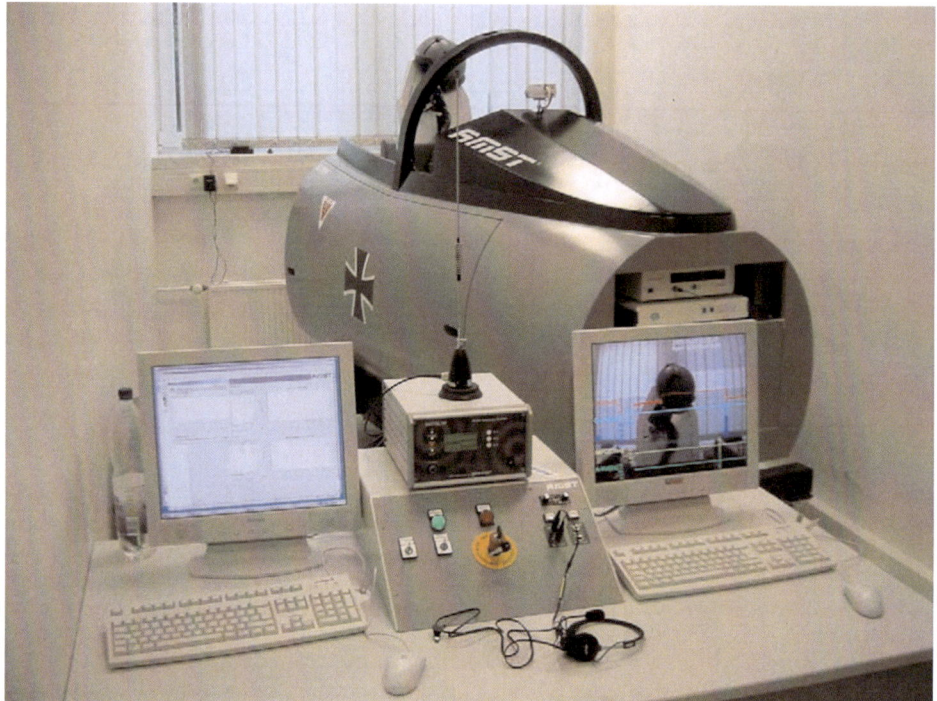

Figure 2.8 The aim of anti-G straining maneuver (AGSM) training is a significant increase in blood pressure to ensure blood circulation to the brain. AGSM trainers like this one help pilots to practice the maneuver in a safe environment. Image courtesy: AMST.

decrease in AGSM endurance after a layoff, although tolerance to peak G for short intervals is less affected. Since practicing the AGSM in an aircraft is an expensive proposition, it's not surprising that AGSM trainers have been developed (Figure 2.8).

The AGSM trainer shown in Figure 2.8 is designed for a generic fighter and features a generic front panel, rudder pedals, side panels, seat, stick, and throttle. Instead of the HUD, a computer screen displays technical and physiological data. The system operates with an instructor sitting in front of the simulator, where he/she is able to select different training modes and modify parameters depending on the type of training. The training consists of the pilot practicing the AGSM components while connected to a medical monitoring system that measures blood pressure (via a finger cuff), electrocardiogram (ECG), breathing frequency, blood flow, chest pressure, and force versus time on the rudder pedals. The objective of the training is to show the pilot how to maintain a high blood level while performing the AGSM. To gauge the pilot's performance, the instructor can view the force exerted on the foot plate and other biomedical parameters and, by using the system's Profile Management option, four training profiles can be created: pedal training, fictive G-load (passive), fictive G-load

Figure 2.9 The author being suited up in the Canadian anti-G suit. Image courtesy: author.

(active), and anti-G profile. The pedal training mode requires the pilot to follow the profiles by pressing the displayed force. The passive fictive G-load mode is defined by a specific G profile designed to teach the pilot the pressure breathing required for an efficient AGSM. While performing this mode, the pilot's G-suit is pressurized according to the G-load generated by the system whereas, during the active fictive G-load mode, the load is increased based on inputs made by the pilot using the stick. The fourth mode – the anti-G profile mode – is performed by the pilot when wearing the full anti-G suit ensemble (pants and vest).

Anti-G suit

As mentioned earlier, anti-G suits were initially developed during the Second World War and changed little over the next 40 years. Today's G-suit (Figure 2.9) generally comprises tightly fitting pants (commonly known as "trousers" in original English), which are usually fitted over the flying suit. The pants are fitted with inflatable bladders which, when pressurized through a G-sensitive valve in the aircraft, press firmly on the abdomen and legs, thus restricting the draining of blood away from the brain during high acceleration.

The current type of G-suit, most of which are inflated by air pressure, is a much lighter variant than the earlier versions, which used water-filled bladders around the lower body and legs. From a human factors perspective, these suits were ideal since the water could be used in a survival situation. Perhaps that's why fluid-filled versions have made a comeback. Recently, the Swiss company Life Support Systems AG and the German Autoflug GmbH collaborated to design the new Libelle liquid-filled suit (Figure 2.10) for use with the Eurofighter Typhoon aircraft.

The Libelle suit depicted in Figure 2.10 comprises pants made of two layers of non-stretch material, with rubber bladders inserted at the calf, thigh, and abdomen. According to pilots who have worn the suit, the system exerts just the right amount of counter-pressure on every part of the pilot's body during acceleration and reacts instantly, even to high G-onset rates. Four double membranes filled with fluids extend from the shoulders to the ankles. Under normal gravity conditions, these membranes are almost flat. As G increases, the hydrostatic pressure increases, causing the membranes to contract. The contraction generated is transferred to a non-stretch material surrounding the pilot's body and effectively stops the blood from pooling in the lower body and extremities. It's a technology that enables the pilot to maintain a high level of awareness and eliminates arm and foot pain at high G.

With so many advantages, you'd assume all fighter pilots wear G-suits, but that isn't the case. Take the Blue Angels, for example. It's a well-known fact in the fighter community that the Blue Angels (Figure 2.11) don't wear G-suits. Are these guys immune to the effects of G? Or perhaps they don't pull many Gs? Well, anyone who's witnessed a Blue Angels demonstration knows that these guys pull a lot of Gs, but they pull Gs in a controlled, gradual-onset

Figure 2.10 The Libelle G-suit is made of a non-stretching fabric and incorporates a system of bladders along the legs and around the torso, filled with a liquid of the same density as blood. Under G, the liquid is forced downwards, the bladders expand, and the suit tightens. Image courtesy: www.flightgear.dk.

Figure 2.11 The Blue Angels is the US Navy's flight demonstration squadron and is the oldest formal flying aerobatic team. The squadron's six demonstration pilots currently fly the F/A-18 Hornet. Image courtesy: United States Navy.

environment. Also, they don't make many high-G maneuvers without periods of rest. Ever wonder what Angels 5 and 6 are doing while 1, 2, 3, and 4 are showing off? They're recovering from the latest G maneuver.

COMBAT EDGE

So we know that by having a proficient AGSM technique and using a G-suit, pilots can buy themselves an extra 3 G of protection but the issue of fatigue remains, which is why the Combined Advanced Technology Enhanced Design G-Ensemble (COMBAT EDGE) was developed. The COMBAT EDGE is designed to help the pilot tolerate sustained Gs as well as assist in proper breathing rhythms for maximum tolerance.

The system, which is the standard for USAF F-15 and F-16 aircrew, is designed to improve tolerance to high-G maneuvers and help prevent G-LOC by reducing fatigue and providing greater endurance. In developing COMBAT EDGE, two factors were identified that could be manipulated to control the blood-pooling phenomenon that occurs during sustained high-G flight. The first factor was the level of pressure applied in the G-suit. Because the cardiovascular system can be likened to an elastic bag filled with blood, the blood moves away from those areas of the bag where pressure is applied. Increase the pressure by squeezing the bag and more blood will be moved. G-suit engineers figured that, by designing G-valves that were sensitive to variations in G-level, higher pressures would be applied to the bladders as Gs increased. The G-valve was important because, in modern fighter jets, the pressurized air source is typically compressor bleed air and the ultimate pressure available far exceeds the pressure needed to counter higher hydrostatic blood-pool levels. Ramping the scale of increasing pressure output relative to the G input defines the performance of the COMBAT EDGE G-valve.

Once the G-valve issue had been resolved, engineers turned their attention to the coverage of the bladders. The pressure being applied to the legs and abdomen was controlled by existing bladders, but the torso was another matter, so the designers tried to figure out how to best control this region. Like the cardiovascular system, the torso can be analogized as behaving as an uncontrolled bag, so why not simply apply pressure to the torso by means of an inflatable pressure vest? The short answer is the lungs. You see, applying pressure to the outside of the chest will squeeze the lung volume making it difficult, if not impossible, for the pilot to breathe. To overcome this problem, COMBAT EDGE utilizes positive pressure breathing (PPB). Here's how it works. When pressure is applied in response to G, PPB is known as pressure breathing for G (PBG). In this mode, the same pressure that is applied to the vest bladders is supplied to the pilot's respiratory system through the oxygen mask. It's a system that has two benefits. First, the pilot can breathe during PBG and, second, since the system is squeezing the pilot's chest from the inside as well as the outside, less blood volume is pooling and is instead being pushed towards the head. It's a win–win solution. Incidentally, because the system supplies positive pressure inside the chest, the COMBAT EDGE vest is referred to as a counter-pressure vest, which we'll discuss here. To ensure correct pressure is supplied to the pilot, the pressure level of the COMBAT EDGE PBG breathing air supply is controlled by the breathing air regulator which, in turn, is controlled by the G-valve. One important system feature is the interlock between the trousers and vest that is created by this method of control, which means the G-valve output pressure that inflates the G-pants is the same pressure that controls the PBG regulator. It's a completely integrated system, which also ensures the G-pants are fully pressurized and inflated before the counter-pressure vest begins to fill and apply pressure (if this level of integration didn't exist and the counter-pressure vest filled first, the blood from the legs and abdomen would be trapped and couldn't rise towards the pilot's head).

Pressure breathing for G

By now, you may be wondering how the PBG system works, so let's discuss that. In the non-PBG mode, the regulator in the breathing air delivery system operates as a demand regulator, similar to the regulator used by scuba-divers. When the pilot switches to PBG mode, the positive pressure flow allows the pilot to inhale and exhale at the designated pressure level. Positive pressure within the mask cavity tends to push the mask away from the pilot's face, causing some leakage, but this isn't an issue as long as the mask cavity pressure is maintained at the controlled PBG level. However, to minimize mask leakage, the COMBAT EDGE system has an inflatable bladder installed in the back of the helmet supplied with the same PBG pressure via a tap in the oxygen mask hose. Since the area of the helmet bladder is greater than the area of the mask cavity, the force pushing the mask away from the pilot's face is overcome when the helmet bladder inflates and applies pressure to the back of the pilot's head. The inflating bladder simply pushes the pilot's face into the mask, tightening the seal, but only during the PBG mode.

Using the PBG system takes some getting used to, so pilots train with the COMBAT EDGE while riding a centrifuge. Once they're accustomed to the strange sensation of PPB in the centrifuge, they practice in the flight environment, conducting simulated aerial combat maneuvers (SACMs) – a highly dynamic environment characterized by a constantly changing G environment. Typical SACM sequences resemble a saw-tooth pattern, alternating rapidly between negative and positive Gs. This is confusing for the cardiovascular system, which doesn't know whether to send blood to the head or the feet but, thanks to the COMBAT EDGE system, pilots have less to worry about than they would if they had to rely solely on G-pants and performing the AGSM. That's because, with the exception of the straining action, the system functions completely automatically. The bladders sequentially inflate and deflate, the mask tightens and loosens, and the pilot can concentrate more on flying the aircraft and less on worrying whether they're about to suffer the dreaded G-LOC. Another advantage of the system is that, at the end of a mission, the pilot who has flown with COMBAT EDGE is much less tired than one flying the same mission with standard life-support equipment.

The only downside to the COMBAT EDGE is the heat load. That's because the counter-pressure vest is an additional layer, which means the pilot sweats more. Before the COMBAT EDGE, there was the Tactical Life Support System (TLSS), an integrated flight suit that had a cooling system composed of tubing threaded inside the bladder layer and filled with cooling liquid. Unfortunately, this liquid cooling system wasn't implemented as part of the original COMBAT EDGE due to logistical reasons. However, when COMBAT EDGE was introduced in Canada, the Defence & Civil Institute of Environmental Medicine (DCIEM) experimented with an air-cooling vest layer positioned beneath the counter-pressure vest bladder. The testing took place in the environmental test chamber at Simon Fraser University (SFU) while I was working there. I remember suiting up subjects in their flight gear and locking them in the chamber for four hours with the temperature cranked up to 40°C. During their stint in the chamber, the subjects reported how hot or cold they felt at 15-minute intervals to assess the effectiveness of the cooling ensemble (the subjects also had a rectal probe

Figure 2.12 STING anti-G suit. Image courtesy: author.

inserted to permit accurate measures of body temperature). More often than not, the subjects complained they were too *cold* – a testament to the effectiveness of the system which is known as STING (Figure 2.12).

The COMBAT EDGE and the STING™ represent state-of-the-art G-suit technology but research still continues to search for even more effective solutions. Common to all approaches that seek to better protect the pilot is the variation of the hydrostatic pressure in relation to the pilot's height. To understand how engineers approach the problem, imagine a pilot sitting in a centrifuge immersed in a cylinder of water with the water at the level of the pilot's chin. The cylinder is attached in a manner allowing the bottom of the cylinder to rotate out as the centrifuge speeds up. Now, as the centrifuge starts to spin and the Gs increase, the direction of G is predominantly downward and, if we measure the pressure of the fluid at various depths, we'll know that the fluid presses on the body at the same pressure as the body fluids are pressing out, trying to pool in the lower extremities. In theory, this should result in a balance of forces, since the elastic bag of body fluids (blood) can't stretch and distend and pool because the bag is restricted from the outside. As the centrifuge spins faster and faster the pressure rises, as does the counter-pressure from the outside – it all happens automatically. Put this into practice, however, and the task is a little more challenging. One attempt at replicating the effect of the column of fluid around the pilot utilized a suit that consisted of a close-fitting bag of water around the body of the subject, thereby replacing the cylinder of fluid. The system was tested only in the centrifuge, where it performed very well, allowing pilots to talk at 9 G and maintain consciousness at 12 G. Unfortunately, the logistical challenges and safety considerations of integrating such a system into a high-performance fighter jet meant the system never got off the ground, but that didn't stop the researchers. Their next approach was to investigate a total-coverage capstan G-suit. The principle of the capstan G-suit is that, as the capstan rotates, it draws the fabric surrounding the limb tighter and tighter. Think of the capstan as a cylindrical bladder running up and down the length of the pilot's limbs. As the Gs increase, the bladders fill with liquid, the pressure inside the bladders increases proportionally, and the bladders distend, thereby exerting pressure on the limbs. It sounds like the perfect system but there are problems. One of the biggest headaches in trying to make the capstan system work is the fit of the liquid-filled bladders. For it to be effective, the system has to be custom-fit and to accommodate the multiplicity of sizes to cater for all pilots would be a logistical nightmare – until recently, that is. With today's technology, pilots can be laser-scanned for measurements of all the anthropo-metric features necessary to fabricate a custom-fitted liquid-filled bladder (LFB) garment.

Of course, liquid can create an additional hazard for the cockpit and, because of the close-fitting requirements, this system wouldn't be compatible with an air-cooling approach, but that doesn't matter so much because the bladder liquid can be temperature-controlled. But what about the signal for the pilots to start their AGSM? In the COMBAT EDGE system, the signal is the start of bladder inflation but, in the LFB suits, this signal is missing because the system performance is so smooth, which means pilots have to be more in tune with maneuvering the aircraft to start their AGSM at the right time.

Compromising G

So far, we've discussed G-LOC and pilot-worn life-support equipment as a means of preventing G-LOC due to G induced by flight maneuvers. But G-tolerance can also be affected by factors other than simply pulling G. Take the night-vision goggle (NVG) for example. You may be wondering how NVGs can affect susceptibility to G, so cast your mind back to the discussion about the PBG system – particularly the problem with the bladder.

When using the PBG, the bladder pushes the pilot's face into the mask (another way of describing this is the helmet being pulled back past the pilot's face by the inflating bladder). Due to the bladder's action, the relationship between the pilot's head and the helmet is altered. This causes problems when NVGs are used. NVGs are mounted on the helmet and are adjusted by the pilots so their eyes are aligned with the exit pupils of the binocular NVG optical system. If the aircraft moves into PBG mode, this alignment may be compromised. Now, as long as the only movement of the helmet is axial, there shouldn't be any reduction in performance, but the main reason for utilizing PBG is to counteract Gs, which usually occurs during dynamic flight maneuvers such as turning the aircraft. In this situation, the force tends to rotate the helmet downward at the front because of the forward CG of the helmet–goggle combination. This is not what the pilot wants because the angular change that results can cause a sideways distortion of the image the pilot is viewing. Of course, if the helmet doesn't slip and shifts only momentarily while the pilot is subject to +Gs, the NVGs will slide back into their original position once the Gs are offloaded. But there is always the risk that some shift will remain once the Gs are offloaded due to the COMBAT EDGE helmet's ability to move on the pilot's head in response to the inflating of the helmet bladder. It is this possibility that makes the fitting of the oxygen mask crucial in maintaining the alignment of the eyes to the NVG images. If the alignment isn't optimized, the pilot must readjust the whole ensemble during flight and that's the last thing a pilot wants while performing aerial combat maneuvers and pulling Gs!

NVGs aren't the only piece of equipment with the potential to compromise G-tolerance. Helmet-mounted displays (HMDs) and helmet-mounted sights (HMSs) can also compromise G-performance. With the HMD and HMS, it isn't so much the movement of the equipment that causes the pilot to be distracted, but the CG displacement that is caused by wearing it. While optical designers and system engineers have sought to reduce the head-borne weight of the HMD and HMS hardware, many of the systems are heavy enough to offset the CG. To reduce the CG impact, engineers have located the hardware away from the front of the helmet, implemented a brow pad fitting system, and modularized the helmet into two parts: an inner helmet known as the life-support module (LSM) and an outer helmet known as the outer mission module (OMM). Together, they are known as the Integrated Mission Helmet (IMH, Figure 2.13), which integrates the HMD and HMS components in a manner that reduces the effects of CG and mask leakage, thereby ensuring a stable and uninterrupted field of view (FoV).

Figure 2.13 Integrated Mission Helmet. Image courtesy: Gentex.

Improving pilot equipment

As with many of the innovations designed to reduce G-LOC, the IMH results in a relatively small improvement in G-tolerance and it is likely that future advances will achieve similar steps forward. For example, further incremental increases in the effectiveness of anti-G-LOC systems may be achieved by the use of body-mounted sensors to continuously relay the pilot's physiological data to the life-support system (LSS). This information would be integrated with the pilot's G history and tolerance limits, thereby enabling an optimized response of the LSS to mission inputs. Taking this a step further, the data could even be used to modify the flight control input to the aircraft so the pilot doesn't exceed their G envelope. In reality, such a strategy is unlikely to be employed given the pilot community's suspicion of anything that comes between them and controlling their aircraft. But that's not to say this information couldn't be useful. For example, G-LOC databases could be developed to allow the aircraft LSS to receive baseline information on the pilot prior to each flight – the data would be stored on a data card, which the pilot would simply slot into the aircraft as part of the pre-flight routine and the LSS would tailor its responses to the pilot. The system could be augmented by body-mounted sensors that would enable the LSS to adjust to any differences between flight datasets and provide the best match between the aircraft capability, the mission, and the pilot.

While the improvements demonstrated by COMBAT EDGE and the IMH consist mainly of alterations to the pilot's equipment, further improvements such as the physiological data card represent a level of integration that will be needed to realize more significant improvements in G-tolerance and reduce the incidence of G-LOC. By now you may be thinking that with all this technology, the pilot would be sufficiently protected from the effects of G, so why continue to search for improvements? Well, in analyzing G-LOC incidents and G-tolerance in general, researchers have learned that what is most important are the last 15 seconds of flight history and whether the pilot has been subject to negative Gs. Why negative? Well, let's consider what negative Gs are first of all. A pilot flying straight and level that pushes the nose of the plane down will experience their weight lessening. The harder the nose is pushed down the more "weightless" the pilot will feel. The effect of these negative Gs is to push the blood into the head – the opposite of positive Gs. But, while the body can tolerate 8 or 9 +Gs without too many consequences, the blood vessels in the pilot's eyes will begin to rupture when as little as 2 or 3 –Gs are applied. This is known as redout (usually preceded by "pinkout") and a pilot who pushes too many negative Gs sees the world through bloodshot eyes, which is not the ideal scenario to conduct aerial combat maneuvers. But redout isn't the worst part of pulling negative Gs. You see, while the pilot is subject to negative Gs, blood is rushing to their head and, because the body senses higher blood pressure in the head, it sends a signal to slow the heart rate down. Of course, a slower pumping rate means less blood is transferred and less oxygen is available to the brain. Now, imagine the pilot pulls out of his negative G maneuver and puts the aircraft into a positive G maneuver. Physiologically, the effect of the positive Gs experienced is multiplied. It's an example of a phenomenon known as "push–pull" and it forms the subject of Chapter 3.

3

The Wobblies

The perils of push–pull

The push–pull effect (PPE) is not a new phenomenon. In 1953, Dr von Beckh conducted research on the effects of transitions from negative to positive Gs and discovered that G-transitions exerted a tremendous effect on the cardiovascular system. In his study, pilots dove from about 3,000 m to about 2,200 m and pulled out of the dive rapidly. The maneuver produced a positive acceleration (hypergravity) of about +6.5 Gz and resulted in several pilots experiencing chest pains, pronounced disorientation, generalized discomfort, and ultimately blacking out. Immediately following the pull-out, the aircraft was flown in a parabola to produce negative G (hypogravity). The results of this maneuver were equally unpleasant, resulting in pilots having to strain harder and blacking out earlier. The results didn't surprise the doctor, who explained the reduced G-tolerance and greater strain as a logical consequence of the transition from hypergravity to hypogravity. There was no record of what the pilots thought of being subjected to the tortuous maneuvers.

Aerobatics

The types of maneuvers investigated by Dr von Beckh are common in the world of aerobatics (Figure 3.1), whose pilots enthusiastically engage in maneuvers with rapid pull-ups and push-overs that exert extreme positive and negative G-loadings. Typical aerobatic competitive sequences may feature outside 360° turns, during which the pilot experiences –2 G for up to half a minute and horizontal rolling 360° turns that produce rapid and repeated oscillations with G excursions varying rapidly between –3 and +4 G and then back to –3 G. Then there is the *outside/inside vertical eight* maneuver, which is perhaps the most physiologically demanding of all. The vertical eight is made up of two loops, one above the other, to form the figure "8". On entry into the vertical eight, the aircraft is usually inverted, with acceleration in the negative 3–3.5-G range. Airspeed decreases during the outside climb, during which more than –3 G are sustained for six or seven seconds. Then, as the airspeed decreases further, acceleration reduces until, at the top of the eight, Gs fall to almost zero. As the airspeed increases on the downward side of the outside loop, the pilot pushes the aircraft under while inverted and the negative Gs rise to more than 5. Then, as the aircraft enters the down leg of the inside lower loop, the airspeed increases

Figure 3.1 The Royal Saudi Air Force Aerobatic Team, Saudi Hawks, flying the British Aerospace BAe Hawk Mk.65. Image courtesy: Royal Saudi Air Force.

together with a positive G-loading. The positive G-loading and airspeed are reduced at the bottom of the lower loop but this phase is rapidly followed by entry to the negative G phase of the climb up to more than –5 G. Then there is a transition from –5 G to +5 G – a difference of about 10 G in less than five seconds. Not surprisingly, such violent and quick G-transitions cause havoc for the cardiovascular system. Heart rates measured during these types of maneuvers have reported changes from 175 beats per minute (BPM) to 40 BPM in less than five seconds! The surges of negative and positive Gs endured during the outside–inside vertical eight are so demanding that only a few of the very best aerobatic pilots are able to pull it off.

Not surprisingly, aerobatic pilots report a full range of +Gz and –Gz effects, with more than 70% experiencing tunnel vision and grayout, 30% reporting blackout, and 20% reporting at least one episode of G-induced loss of consciousness (G-LOC). Other occupational hazards include skin rashes, back, neck and arm pain, headaches, difficult breathing, and even urinary incontinence! While these G-effects, which are primarily related to positive G, are worrying enough, even more dangerous are the PPE-influenced effects induced by negative G. For example, many advanced aerobatic pilots experience bloodshot eyes and redout due to the gravitational effects of negative G. Other negative G-effects include bright scotoma, red scotoma, eyelid rashes, eye pain, and persistent vertigo, also known as the *wobblies*.

The wobblies

This latter symptom – despite its innocuous name – is perhaps the most serious of all the PPE-related effects, since it can be near incapacitating in flight and may persist for days and even weeks after onset. The wobblies, more commonly known among flight surgeons as G-induced vestibular dysfunction (GVID), is very similar to benign paroxysmal positional vertigo (BPPV), which is caused by inner-ear dysfunction in which crystals are thought to migrate into the posterior semicircular canal, producing a false sense of angular motion and consequent vertigo. The most likely cause of the migrating crystals is the angular acceleration that aerobatic pilots expose themselves to. Angular acceleration occurs when there is a change in velocity. In the wobblies, the crystals are forced into the posterior semicircular canals by the shear forces induced by positive and negative G maneuvers. A pilot who suffers from the wobblies usually experiences a whirling sensation in which the horizon appears to spin. There is no loss of consciousness, but pilots usually feel nauseous and, when they leave the aircraft, they will often lean against it or immediately sit down.

Aerobatic pilots aren't the only ones who suffer from PPE-related effects. Fighter pilots engaged in aerial combat maneuvers (ACMs) are involved in extremely dynamic movements that are very similar to the maneuvers that aerobatic pilots perform. An ACM maneuver begins when a pilot realizes he is under attack. His first priority is survival, so his opening moves are defensive, with the attacker conforming predictably to the defender's movements. While each maneuver has its counter, it is the precision and timing of a maneuver that are important: the ability to out-fly an opponent. To be successful in outwitting his opponent, a pilot has to prevent the attacker from positioning in the lethal or vulnerability cones, which means being able to do some fancy flying. This is why aspiring fighter pilots are taught defensive and offensive ACMs such as the Break, the Scissors, the High-G Barrel Roll, Junking, the Spiral Dive, the Vertical Rolling Scissors, the Split S, the High Speed Yoyo, the Vector Roll or Rollaway, the Lag Pursuit, the Low Speed Yoyo, the Barrel Roll Attack, the Vertical Reverse, the Immelmann, and various versions of and counters to these.

Let's take a look at a couple of these. We'll start with the High-G Barrel Roll (Figure 3.2). This maneuver is used against an attacker closing fast from astern. It starts with a break, then a roll in the direction of the break. If the attacker is closing fast and is caught by surprise, he may easily fly through and end up in front, thereby reversing the positions. If he attempts to follow the barrel roll, he will probably end up high and wide of the defender, who can then turn in towards him, forcing him down and in front.

The High-G Barrel Roll is a difficult maneuver to execute successfully, and is quite easy for the attacker to counter. A more difficult maneuver is the Rollaway (Figure 3.3). In this maneuver, the attacker reaches the top of the yoyo and rolls in the opposite direction to the defender's turn, which has the effect of pulling him tighter behind the defender.

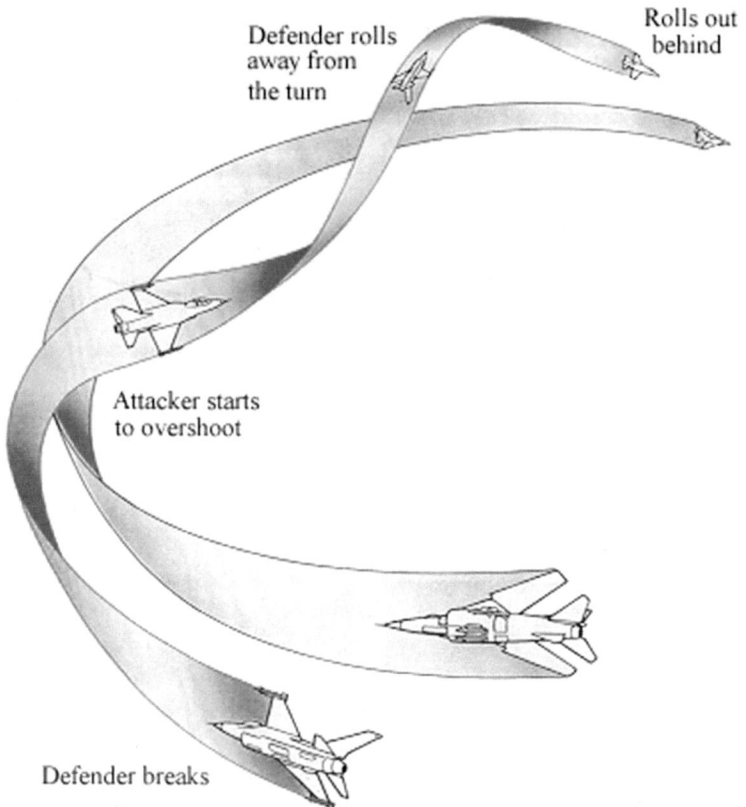

Defender rolls
away from
the turn

Rolls out
behind

Attacker starts
to overshoot

Defender breaks

Figure 3.2 The High-G Barrel Roll is a difficult maneuver to execute successfully, and is in fact easy for the attacker to counter. It will only work if the attacker has been led into, or is in, a high angle-off, high overtake situation. Image courtesy: Wikimedia Commons.

The High Speed Yoyo maneuver is performed when the attacker realizes he is unable to stay on the inside of the defender's turn. The attacker relaxes his angle of bank and pulls high. As he comes over the top, he is inverted, looking down at his opponent. His speed falls during the climb and this reduces his turn radius, which is reduced even further when he turns in the vertical plane, but, by doing this, the attacker is well positioned to slide into a firing position. A perfectly executed High Speed Yoyo is fiendishly difficult to counter but there are a few tricks that the defending pilot can try. If his energy state is high enough, he can pull up into the attack, but that would risk depleting his energy reserves to the point where he can no longer effectively defend himself. Alternatively, as the attacker pulls up his nose, the defender can relax his turn and spiral wide at full throttle, which might increase his speed and widen the distance. But, if the attacker has misjudged his maneuver and rolls out close astern but high, all the

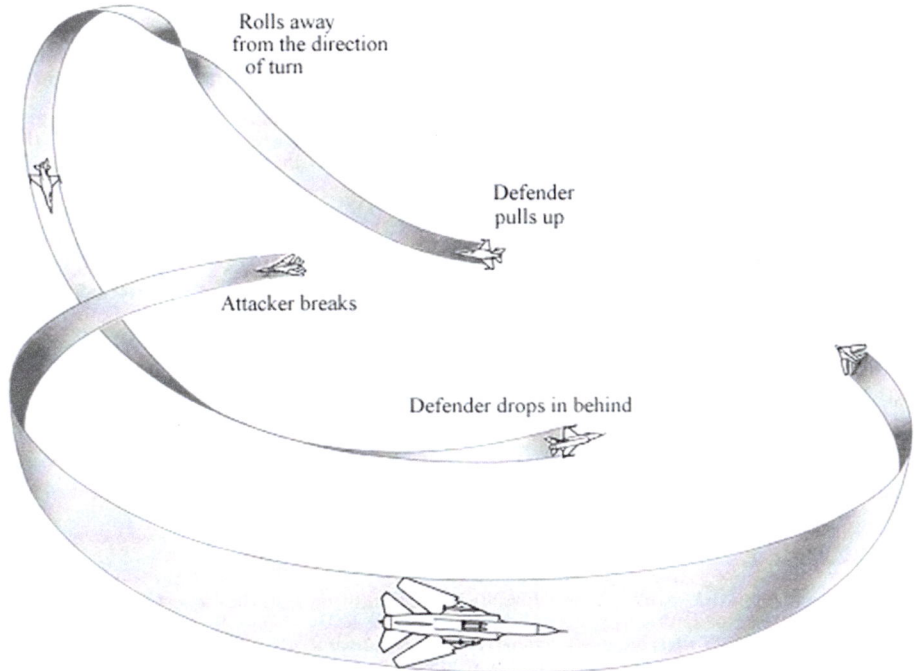

Figure 3.3 The Rollaway maneuver is used against an attacker closing fast from astern. It starts with a break, then a roll in the opposite direction to the break. Because it's a high-G maneuver, a lot of speed is lost, up to 100 knots in some cases, particularly if performed "over the top". Image courtesy: Wikimedia Commons.

defender has to do is relax his turn to maintain speed. Then, when the attacker drops his nose and dives, the defender is able to reverse hard up into him.

As you can imagine, such rapid movement through all the spatial planes with the resultant rapid changes in gravitational force vectors presents a very challenging environment not much different from that which aerobatic pilots have to contend with. Also, the high transient G exposures that comprise many of the ACMs cause similar PPE-related problems, which is why scientists continue to investigate the phenomenon.

Push–pull research

Forty years after Dr von Beckh's work, more research attempts were made investigating the hypergravity-to-hypogravity G-transition. For this research, scientists based their observations on the outcome of inflight tests and experiments conducted using a modified centrifuge known as the Coriolis Acceleration Platform (CAP), a device capable of generating negative and positive

Gs thanks to its unique capability of combining angular and linear motion. The subject carrier was a 1.2-m-long capsule mounted to a movable platform. Alterations in the subjects' negative and positive accelerations were achieved by placing each subject in a supine seated position and moving the capsule linearly along the track while the whole device rotated. Of course, as the CAP rotated, the direction of acceleration reversed, which meant that subjects were subjected to positive G when rotated "feet-out" and negative G when rotated "head-out". The magnitude of the G depended on the distance of the capsule from the center of rotation.

Tests were monitored by a two-way intercom and video monitor so researchers could rapidly identify distress and/or loss of consciousness. Peripheral light loss (PLL) was assessed using a light bar system comprising a central flashing red light and four flashing green lights at the subjects' eye level. This way, subjects could report to the researchers how the Gs were affecting their vision. Each subject underwent a series of five Gz transitions ranging from –2 to +2.25 Gz (G onset rate was 1 G/second) separated by 30-second rest periods while the CAP rotated at +1 Gz. The five Gz transitions included two control transitions that exposed the subject to +1 Gz for at least 30 seconds, followed by +2.25 Gz for 15 seconds before a return to +1 Gz. The three experimental transitions exposed the subject to –2 Gz for 2, 5, and 15 seconds followed by +2.25 Gz for 15 seconds before returning to +1 Gz. After analyzing the transitions, the researchers found that tolerance to +Gz following exposure to –Gz was affected by the duration of the preceding –Gz exposure. They also found that –Gz exposures as short as two seconds were enough to have a negative impact on +Gz tolerance, making PPE a serious flight concern. Now, the results of the CAP study were based on responses by pilots riding a centrifuge, which meant there were potential confounding variables prior to the onset of increased acceleration. That's because G-transition studies that use centrifuges produce what are known as G-vector biases and Coriolis cross-coupling. The Coriolis Effect and cross-coupling refer to the vestibular effect of tilting the head during whole-body rotation. For example, the Coriolis Effect can occur in flight when the pilot rotates their head about one axis while the aircraft is rotating about another axis (very discombobulating), with the result that the semicircular canals receive conflicting sensory information, causing a disorienting sensation. Now, imagine the pilot in a centrifuge. To provide –Gz or +Gz exposure, the centrifuge generates simultaneous rotation about two axes – the *angular* acceleration of the gondola at the end of the swing arm and the *linear* acceleration of the gondola spinning. This rotation effect generates vestibular stimulation that can produce changes in blood pressure, which can in turn confound data.

To understand better how PPE may affect a pilot flying an aircraft rather than one sitting in a centrifuge, it's helpful to imagine what a pilot does to produce negative and positive Gs. The PPE is so named because it describes the stick action inputs of the pilot when performing an aerobatic maneuver. Pushing the stick forward develops negative Gs while pulling back on the stick results in positive Gs. As we know from the CAP studies, these two G situations have physiologically opposite responses so, when the pilot starts the +G period, the starting position is much farther from what is needed to compensate for the +G experience and it takes much longer for the body to recover. This causes a problem in terms of G-tolerance. For example, a pilot exposed to just 2 negative Gs who is immediately re-exposed to positive G will find their G-tolerance reduced by about 40%. Also, the greater the exposure to negative G (either in magnitude or duration), the greater the effect on the pilot's ability to tolerate positive Gs. Not surprisingly, PPE has been the primary cause in several G-LOC-related accidents, two of which are described here.

> *November 1977.* The accident aircraft was one of two Canadian Forces (CF) 104s that were performing ACMs. The student pilot occupied the rear seat of CF104D, which was normal procedure for ACM training. After the set-up at 25,000 feet, the student initiated a planned negative G defensive maneuver, the second of the sortie. As planned, the aircraft executed the –2 G maneuver for several seconds, and then extended at relative negative G for several more seconds. The aircraft then flew a slice-back to about +4 Gz. During that maneuver, the aircraft entered a descent with approximately 135° of bank and 60° nose pitch down. When the aircraft entered cloud at approximately 4,000 feet, the dive had decreased to 20–30° nose down and 45° bank. Although a radio call to "knock-it-off" was made as the accident aircraft passed 16,000 feet, there were no radio transmissions heard from the mishap aircraft. On impact, the wings were level, pitch attitude was slightly nose-up, and speed brakes deployed. Both pilots were killed on impact.

> *July 1995.* During an ACM mission at Cold Lake, Canada, the accident pilot experienced relative negative Gs for approximately eight seconds during a head-on pass with the second aircraft. He then commenced a nose-down slicing maneuver prior to Gz loading to a plateau of +5.6 G. After about five seconds, the pilot began to ease off the turn rate (relaxing the G), continued the roll, and increased the angle of descent to >70°. The pilot initiated a recovery attempt at extremely low altitude with insufficient time to pull out. The aircraft impacted the ground at Mach. The pilot did not attempt to eject and was killed.

Physiology of push–pull

The above examples are just two of a number of accidents that have been attributed to PPE. In fact, due to the dangers of PPE, some air forces have prohibited certain –Gz maneuvers, but there are some situations in which negative-to-positive Gz transitions are inevitable in flight. For example, in the world of the jet fighter pilot, there are certain maneuvers that are used to regain energy to acquire tactical advantages over an opponent. One such maneuver involves unloading the aircraft from positive-to-negative Gz prior to accelerating. This type of maneuver involves rolling and pulling the maximum G-forces available, resulting in the dreaded positive-to-negative G-transition. But why is the transition from negative to positive G so dangerous and why did those subjects riding the CAP lose so much of their G-tolerance? The answer can be found in cardiovascular physiology. When a pilot experiences –Gs, there is a *decrease* in heart rate, or bradycardia, and when the pilot is exposed to +Gs, there is an *increase* in heart rate, or tachycardia. The bradycardia takes about two to four seconds of –Gz exposure to develop and, with increased –Gz exposure, there can be some recovery but, since the effects of –Gz are so unpleasant, most pilots tend to avoid it (aerobatic pilots are an exception). Although the effects of –Gs are uncomfortable, the most dangerous aspect isn't the discomfort, but the timing. Here's why. Tachycardia takes about six to eight seconds of +Gz exposure to develop and the rate of bradycardia development is faster than the rate of tachycardia development. Now, the brain has a latency period of about five to seven seconds during which it can continue to function without blood flow. But, the recovery time of six to eight seconds, which is the time for tachycardia to compensate for the +Gz load, is greater than this latency period, which means that, if full –Gz reaction occurs and then +Gz is experienced, the brain doesn't get the blood it needs to last through the recovery time. The end result is G-LOC.

Another mechanism known to contribute to the PPE is something physiologists refer to as "peripheral vascular contractility changes". This a measure of how flexible the cardiovascular system is. Research has shown that the rate of change of blood vessels relaxing (vasodilation) is faster than the rate of change of them tightening (vasoconstriction). This is of concern to pilots because –Gz causes vasodilation and +Gz causes vasoconstriction so, again, there is that latency period that means the recovery is slower than the initial reaction. An interesting correlation is that, in older pilots whose blood vessels are less flexible, there is a reduction in the PPE because the initial compensatory responses to –Gz don't occur on the same time scale, which allows the body to react to the changes more effectively.

Avoiding push–pull

Despite all the latest technology and research into the causes of negative and positive Gs, PPE continues to be a problem for pilots, whether they're a fighter pilot flying the latest F-22 or an aerobatic pilot pulling Gs in a Pitts S-2 Special. So, what can a pilot do to minimize the risks of PPE?

Well, first of all, pilots can monitor their fatigue, since this severely degrades a pilot's ability to perform the anti-G straining maneuver (AGSM) and also affects alertness and G-awareness. Either way, G-tolerance is compromised. Next, it's important for pilots to avoid heat stress as much as possible, as this degrades the body's ability to do work and reduces G-tolerance. The combination of dehydration and blood moving to the skin for cooling significantly reduces G-tolerance and work capacity to such a degree that a pilot who has lost just 3% of their bodyweight to dehydration may experience a 50% reduction in G-tolerance. Equally important is the role of nutrition since, when blood sugar drops, so does G-tolerance. Then there's the effect of alcohol and its hangover effect which, not surprisingly, has been well documented to have a significant negative impact on G-tolerance. But that's not all. Alcohol also degrades sleep quality, causes dehydration and salt loss, depletes body sugar stores, and dilates blood vessels – all factors that can have a negative effect on a pilot's ability to tolerate G-stress.

Desdemona

In William Shakespeare's play *Othello*, Desdemona is a Venetian beauty who enrages and disappoints her father, a Venetian senator, when she elopes with Othello, a man several years her senior. Desdemona also happens to be the acronym for a multi-purpose disorientation simulator (Figure 3.4) that takes its name from Desoriëntatie DEMONstrator – whether the intent was to name the simulator after Shakespeare's character is unknown. What is known is that the simulator is the result of the collaboration between TNO (Toegepast Natuurwetenschappelijk Onderzoek, which translates to Netherlands Organization for

Figure 3.4 AMST's Desdemona® is a unique simulation tool, due to the special characteristics of its innovative motion platform. Image courtesy: AMST.

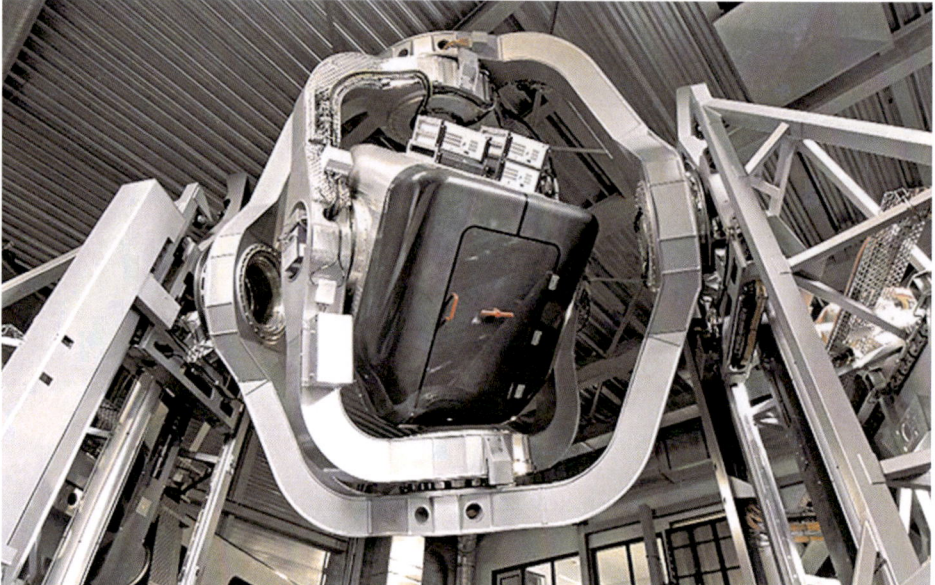

Figure 3.5 The Desdemona® rotating gimbal system. Image courtesy: AMST.

Applied Scientific Research) and AMST (an acronym derived from Austria Metall Systemtechnik), an Austrian company that also happens to be in the centrifuge-building business.

AMST was responsible for the design of the rather intimidating and unique simulator (it's the only one in the world). The company also installed it at TNO's location in Soesterberg, The Netherlands. The simulator features a unique non-synergistic motion system with six degrees of freedom (DoF), which gives the system a broad range of dynamic performance capabilities similar to those found in standard dynamic flight simulators. But, the rotating gimbal system (Figure 3.5) allows the Desdemona motion system to replicate unusual attitudes and large attitude changes one-to-one.

The key element in the system is the cockpit, which is fully gimbaled, allowing for unlimited rotation in all directions. The pilot, with their head positioned in the center of rotation, controls the flight instruments and has a view outside via up-to-date visuals. For simulation purposes, different aircraft models and databases are available. The versatile design allows the cockpit to move along an 8-m horizontal track vertically over 2 m (horizontal and vertical accelerations allow for 0.5 g). Since the horizontal track is rotated around a vertical axis, the system can work as a centrifuge, capable of simulating all manner of maneuvers and sustained G-loadings. Whereas the working principle of a normal centrifuge is variation of the angular velocity with constant eccentricity, resulting in a varying G-load, Desdemona keeps the angular velocity constant and varies the eccentricity. Generating G this way means the onset of a sustained G isn't always accompanied by a strong angular acceleration, as is the case in a conventional centrifuge. This is because it's possible to rotate the system with the cockpit in the center position before moving it away from the axis without varying the angular velocity. The second unique feature of the system is that the cockpit can be moved in the vertical direction, which allows variations in Gx or Gy during a sustained Gz load to be generated almost without concurrent variation of the angular motion of the cockpit – this is a boon for pilots wanting to experience motion perception during post-stall maneuvering.

The modular and software design of the fully gimbaled cabin enables fast reconfiguration of the interior and the installation of human machine interfaces, all of which makes for a versatile research system that is ideal for studying PPE effects. As you can see in Figure 3.4, unlike conventional motion systems, the Desdemona motion system is not a synergistic or parallel robotic system. Instead, it has a non-synergistic or a serial motion system with six axes that can be moved independently. The six axes are central yaw, radius (R), heave (H), cabin roll, cabin yaw, and cabin pitch. Thanks to the cabin being suspended in a fully gimbaled 3 DoF system, the configuration allows unlimited cabin rotation around any arbitrary axis in space. The gimbaled system is mounted in a heave system that translates the gimbal system and the cabin in the vertical plane. The heave/gimbal system can be moved horizontally over a sled that you can see in the image. Not only that, but the sled can be rotated unlimited in the middle around a vertical axis and the central yaw axis: a combination that gives the Desdemona the capability of generating a sustained 3-G load. Since the system is designed for research, it's not surprising the simulator is loaded with instruments and measurement systems. The first measurement system comprises the various position encoders mounted on each axis used by the electrical drives to control each axis position. The second measurement system comprises three solid-state accelerometers mounted on the cabin chair at the position of the subject's head. These three sensors are intended for safety purposes to measure the force vector exerted on the subject. The third measurement system is an Inertial Measurement Unit (IMU) comprising a fiber-optic and temperature-compensated three-axis gyro in combination with a solid-state three-axis accelerometer. The gyro is rigidly attached to the inside of the cabin structure and the accelerometer is also attached to the cabin but with a flexible mechanical connection. This provides digital output signals of the three cabin accelerations and the angular rates. In dynamic flight simulator mode, with the cabin fixed at the maximum radius position of 4 m, the central yaw axis velocity and acceleration can be used in combination with the cabin orientation to generate a 3-G force. The three-axis gimbal system combined with the unique capability to rotate the cabin 360° allows the Desdemona to simulate not only negative and positive G-transitions, but also other aerobatic maneuvers such as barrel rolls, a maneuver that is next to impossible to simulate with traditional motion systems.

4

The G Machine

Riding the disorient express

"We try to make this training so intense and so ingrained into their body and their muscle memory that it comes as second nature to them. Because if they're doing all these other maneuvers ... and they start pulling Gs, the possibility of them G-LOC'ing and people dying and losing aircraft just increases exponentially."

Tech Sergeant Carlos Rivera, centrifuge instructor,
Holloman Air Force Base

I'm strapped into a modified F-15 fighter jet seat (Figure 4.1) inside a gondola (Figure 4.2) half the size of a Smart Car with my feet on a force plate and my eyes

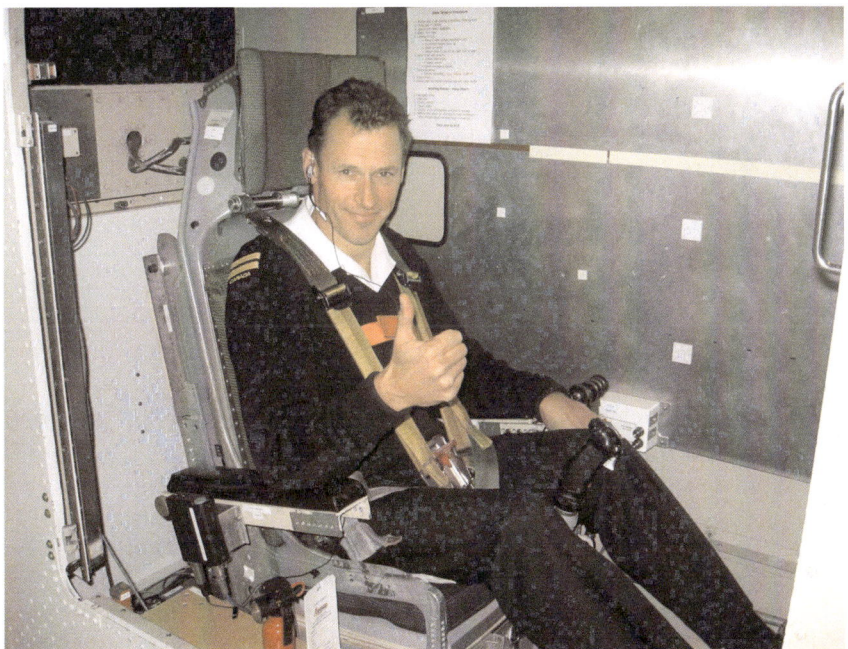

Figure 4.1 The author ready for a centrifuge run in Canada's only human centrifuge. Image courtesy: Chris Townson.

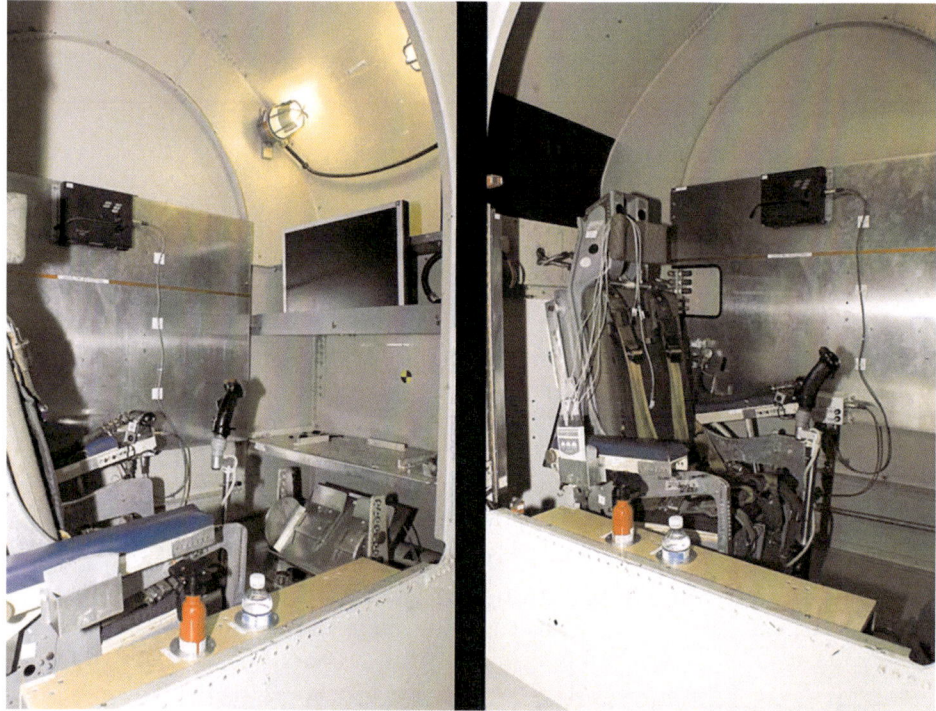

Figure 4.2 DRDC centrifuge gondola. Image courtesy: Jim Clark, DRDC.

staring straight ahead at an X-Box-style representation of a chase plane. The gondola is attached to one end of the 6-m swing arm of Canada's only human centrifuge (Figure 4.3), a $20 million piece of hardware located in Building 54 at Defence Research Development Canada Toronto (DRDC-T). Below me in the "Pit" is a 300-horsepower engine that causes the entire apparatus to spin on an axis inside a 19-m-wide circular bunker.

Spinning the G machine

"Okay, so we'll just run through our checklist here. Operator. Confirm two DVDs are recording." That's the voice of Chris, one of our Acceleration Training Officer's (ATOs), seated in the control room, aka Mission Control (Figure 4.4), going through his checks, one of which is to make sure there is a video record of my attempts at performing a proficient anti-G-straining maneuver (AGSM). The ATO runs the show, giving instructions to Allison, the Operator seated next to him, to launch and terminate runs. Being an ATO is an unusual job simply because there aren't that many human centrifuges (fuges) in the world. The National Aerospace Training Center (NASTAR) in Southampton, Pennsylvania, is a commercial "fuge", NASA has a couple, and the United States Air Force (USAF) has a couple

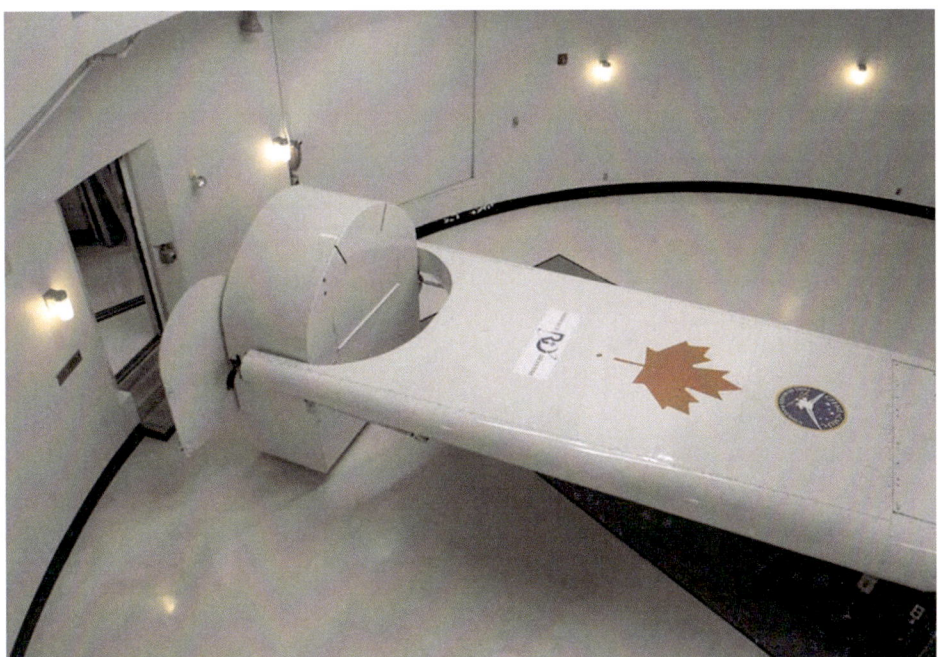

Figure 4.3 DRDC centrifuge can generate 1.4–15 G, at onset rates of up to 2.8 G/second, although this centrifuge has been limited to 10 G. The free-swinging gondola, which takes payloads of up to 320 kg, is equipped with an aircraft-type seat and pressurized air for breathing or life-support equipment, a rider-operated enable switch, and two-way audio communication. Image courtesy: George Kourounis ©.

more. Outside North America, centrifuges are equally rare. Poland, Sweden, Austria, the United Kingdom, The Netherlands, and Germany each has one, as do Singapore and China. All told, the number of centrifuges and ATOs globally is no more than a couple of dozen.

"Confirm two DVDs recording." Allison confirms the recording and, anticipating the next action, switches to the screen showing the G profile.

"Erik. How do you read?" Chris asks. That's the cue for me to confirm I can hear Chris on the communication loop.

"Five by five" I respond. I'm sweating already. Not because I'm nervous, but because it gets so damn hot inside the gondola and the only ventilation comes from a pitifully underpowered fan 30 cm from my face.

"Operator. How do you read?" Chris asks.

"Operator five square" comes the response.

"Operator. Check AGOR is on the board." That's Chris asking Allison to confirm that the first profile is loaded. The ATO sees exactly what the Operator sees but this is the military, so redundancy is everything. The first profile – the Acceleration Gradual Onset Rate (AGOR) profile (Figure 4.5) – is one that

Figure 4.4 DRDC centrifuge control room. Image courtesy: George Kourounis ©.

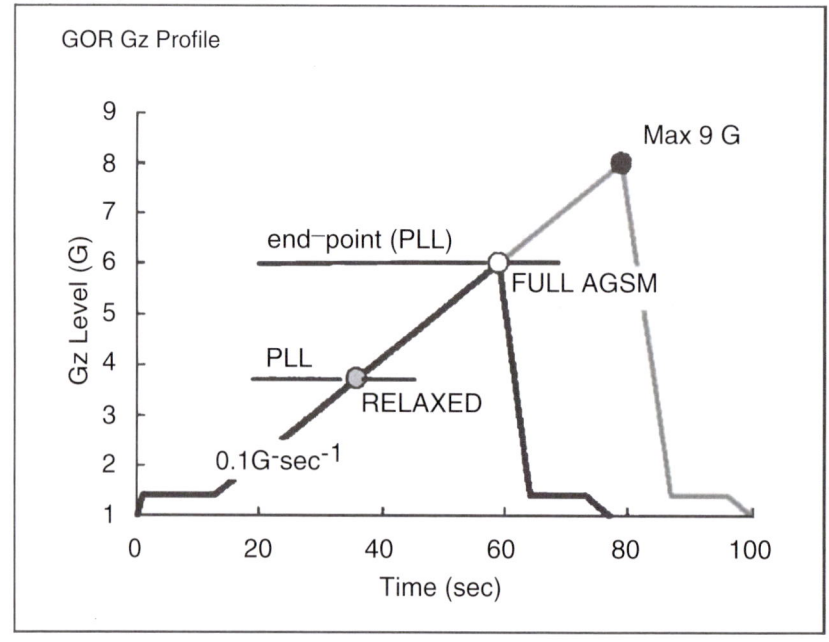

Figure 4.5 Gradual-onset profile. Image courtesy: DRDC.

increases the Gs at a gradual onset rate (GOR) of 0.1 G/second from the fuge's idle speed of 1.4 G.

"AGOR on the board" Allison confirms.

"Check AGOR on the board. OK, Erik. Are you ready to go to baseline?" Chris asks.

"Ready." I grip the control stick, ensuring the enable switch (a kill switch I can release at any time to bring the fuge to a full stop) is fully depressed.

"Operator, are you ready?" Chris asks.

"Operator ready."

"OK. Operator. Launch centrifuge to baseline" Chris commands.

"Launching centrifuge to baseline in . . . three . . . two . . . one . . . launch." In Mission Control, Allison launches the centrifuge and the engine down in the Pit starts to do its work. Inside the gondola, the start of the fuge spinning up is almost imperceptible but, as the machine picks up speed, the gondola slowly swings out (Figure 4.6) and there is the sensation of movement. I stare at the X-Box chase plane ahead of me on the screen.

"OK Erik, you're going to feel a slight pitch-up sensation as the centrifuge starts to spin up. Just let us know when that sensation has gone away." Chris's advice is

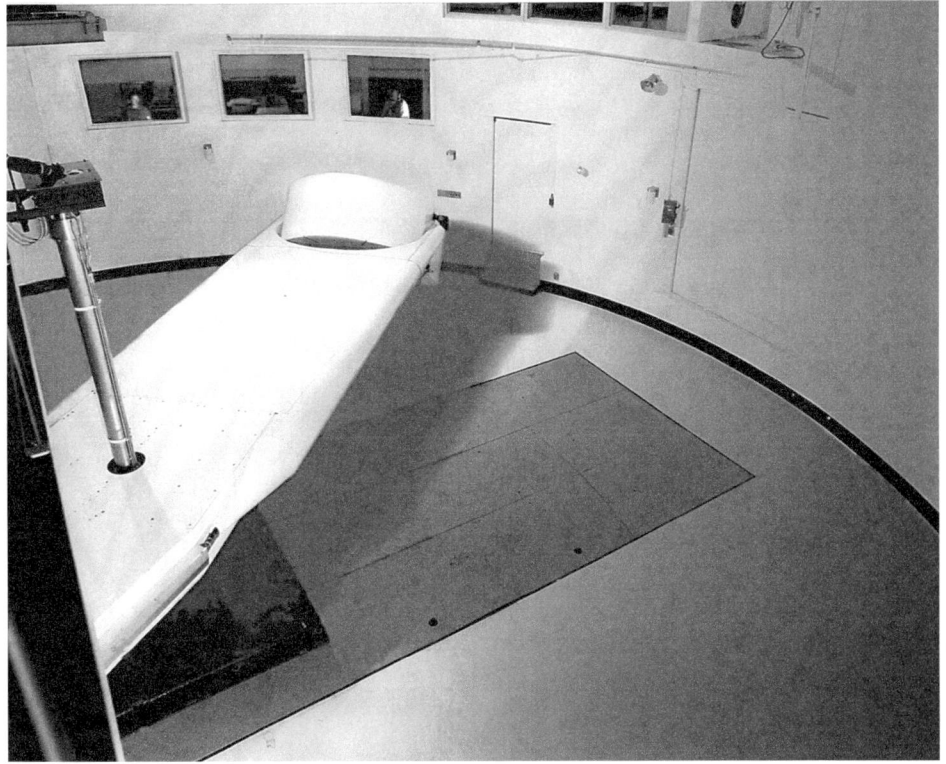

Figure 4.6 Gondola swinging out as it spins up to idle. Image courtesy: DRDC.

familiar because it's the same advice I've given to hundreds of pilots over the years.

"*Idle.*" Allison confirms the fuge has reached its idle spin rate of 1.4 G. It's now ready to launch the first profile.

"*Check idle*" Chris confirms.

"*The pitch-up has gone*" I confirm.

"*Check. Are you ready for the profile?*" Chris asks.

"*Yes, I'm ready.*"

"*Operator, are you ready?*"

"*Operator ready with AGOR on the board*" Allison replies.

"*Check AGOR on the board. Operator, launch AGOR*" Chris says.

"*Alright Erik. Just sit back and relax. Launching AGOR in . . . three . . . two . . . one . . . launch*" Allison says.

The run begins and the centrifuge starts to accelerate at one-tenth of a G per second.

"*2 G*" Allison announces.

"*Check 2 G*" comes Chris's reply.

3 G goes by and I begin to be pushed into the seat. 4 G. Breathing requires noticeably more effort now. An arm's length from my eyes is a metal strip known as a light bar. At each end is a green light and in the center is a red light. As 4.5 G is announced, the two green dots begin to march inward, starting from just beyond my peripheral vision. It's the sign to start my AGSM. I take a deep breath and simultaneously apply force to the footplate, tensing my leg and abdominal muscles. Immediately, my vision expands to normal. Inside Mission Control, Chris has a light bar indicating how much force I'm applying.

"*5 G*" Allison announces matter-of-factly. For the hundreds of people who tour the DRDC-T facilities every year, seeing the fuge in action is the highlight of their day. But, for the fuge staff, it's business as usual.

"*Check 5*" comes Chris's response.

I complete my fourth AGSM cycle as 5.5 G is announced. Thanks to performing an efficient AGSM, my vision is crystal clear. No grayout or loss of peripheral vision. But cranking out the AGSMs is hard work.

"*6 G.*"

The centrifuge is spinning at almost a revolution a second at close to 60 km/hr and I weigh a G-induced 360 kg. The centrifuge continues for a couple more seconds before I hear the order to release the enable switch.

"*Terminate, terminate, terminate. Keep straining on the way down*" Chris says.

I immediately release the enable. The kill switch does its job and the fuge starts spinning down, precipitating a tumbling sensation that feels as if I'm falling down a mine-shaft – not that different from riding the Desperado rollercoaster. I continue to perform the AGSM, mindful that it's still possible to lose consciousness as the fuge spins down to idle.

"*Idle*" Allison announces.

"*Check idle*" Chris confirms.

The fuge is back at 1.4 G and everything is back to normal. After a brief

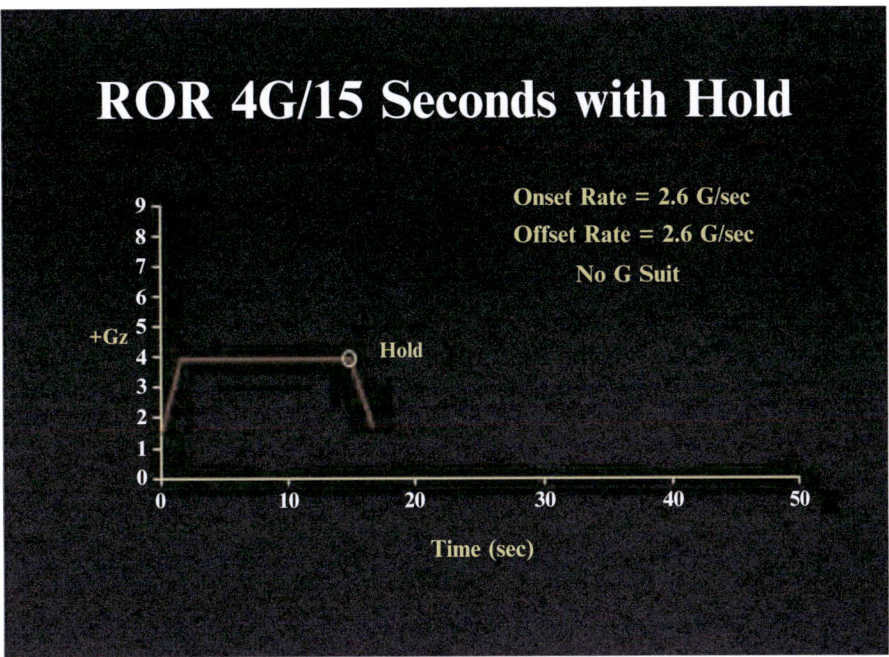

Figure 4.7 Rapid-onset rate profile to 4 G. Image courtesy: DRDC.

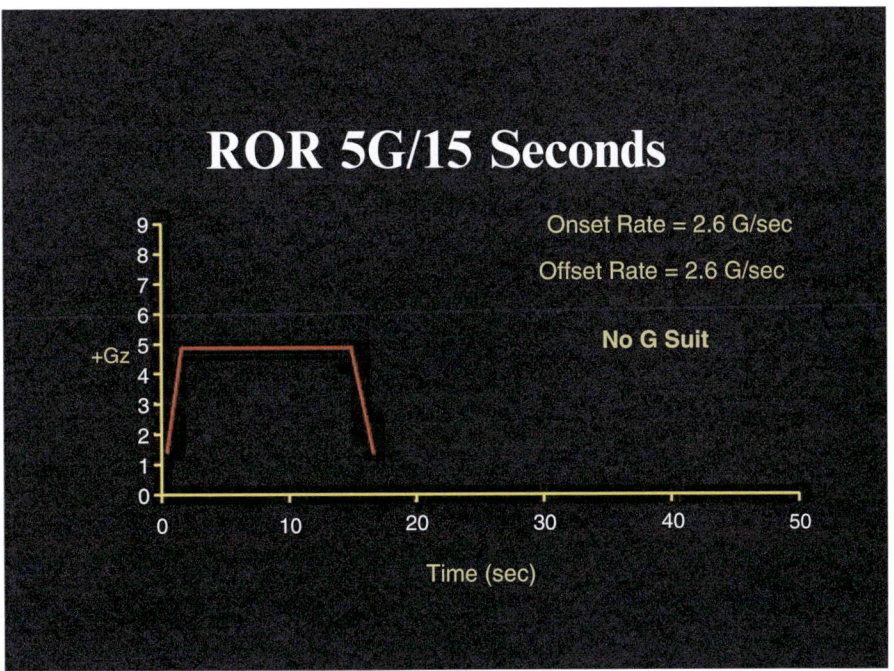

Figure 4.8 Rapid-onset rate profile to 5 G. Image courtesy: DRDC.

recovery, Allison loads a rapid-onset rate (ROR) profile to 4 G (Figure 4.7), which I complete before going on to finish the ROR 5 (Figure 4.8).

A typical human centrifuge like the one I've just described consists of a horizontally aligned arm with a cabin/gondola for the passenger/rider on one end. The arm rotates around a vertical axis and is powered by an electric motor. A training centrifuge like the one at DRDC-T is used to increase accelerative force under controlled conditions. It prepares astronauts and fighter pilots for the G-forces they will experience during launch or re-entry or in a fighter jet. The simulation is used not only to increase the rider's "G-tolerance" and allow them to practice the AGSM, but also to determine what physiological effects can result from exposure to increased accelerative forces.

The origins of centrifugation date back to the beginning of the nineteenth century. Back in those days, rotating a person by placing them along the arm of the centrifuge was believed to be conducive to treating nervous and mental diseases. The first modern human centrifuges were not built until the 1930s. These devices could produce accelerations of up to 20 G, which is still the maximum acceleration that training centrifuges reach today. During the Second World War more centrifuges were built and developed, but centrifugation didn't become widely used for training until much later when incidents of gravity-induced loss of consciousness (GLOC) were reported to have caused fatal aircraft accidents in the 1970s. In the 1980s, in an attempt to reduce the risk of GLOC-related incidents, air force training facilities in many countries accepted centrifugation as a regular part of pilot and astronaut training. Which brings us to the subject of manned spaceflight and the Johnsville centrifuge.

Johnsville

Data obtained from the first Soviet and American satellites of the late 1950s indicated that the atmosphere extended much higher than previously thought. Until the data from these satellites were analyzed, scientists had assumed that deceleration re-entry forces (eyeballs out) would be greater than the acceleration of launch (eyeballs in). Based on this assumption, military physiologists working at the Johnsville centrifuge exposed chimpanzees to a peak of 40 G for one minute. Post-run examinations weren't very encouraging, since the primates had internal injuries and signs of heart malfunction. Back in the late 1950s and 1960s, when centrifuges made their debut as a test of astronaut selection, astronauts didn't have many positive things to say about the device. Even American hero, John Glenn, who wasn't scared of anything, called the centrifuge a "dreaded" and "sadistic" part of astronaut training. Apollo 11's Michael Collins called it "diabolical". *Time* magazine referred to it as a "monstrous apparatus" and a "gruesome merry-go-round". The object of the astronauts' bête noire was the Johnsville human centrifuge (Figure 4.9), the machine future spacefarers loved to hate.

The behemoth depicted in Figure 4.9 was operated by the Navy at its Naval Air

Figure 4.9 The Johnsville centrifuge. Located at the Aviation Medical Acceleration Laboratory of the Navy Air Development Center, Johnsville, Pennsylvania, the centrifuge was used throughout the early American space program. With a 17-m radius, a rate change of 10 G/second and a maximum capability of 40 G/second, it was the most powerful centrifuge then in existence. Prior to the arrival of the Mercury astronauts in August of 1959, the Johnsville centrifuge was used for testing by X-15 pilots, including Neil Armstrong prior to his selection by NASA. Image courtesy: United States Navy.

Development Center in Warminster, Pennsylvania, just outside Philadelphia. The centrifuge resided in a 1,022-m^2 cylindrical building and consisted of a 3-m-diameter steel gondola attached to the end of a 15-m arm driven by a monster 4,000-horsepower motor. Not only did the gondola whip around the room, it could rotate as well, which allowed the astronauts to experience a fairly close simulation of the G-forces they'd likely face as they launched and re-entered the atmosphere. If the ball spun one way, it generated positive Gs, while spinning it the other way generated negative Gs. In 1963, the original gondola was replaced with a more spacious three-seat model that was used for training Apollo crews. Since centrifuge research was still in its infancy, every astronaut who was strapped into the Johnsville fuge was essentially a guinea pig, providing researchers with valuable insight into what the human body could and could not withstand. In fact, much of the Mercury training was conducted just to learn what happens if humans were exposed to high accelerations in different vectors. To that end, astronauts were subjected to re-entry profiles, take-off profiles, and even long-term exposures. The astronauts spent so much time in the "wheel" that it soon became a rite of passage. One objective of such ordeals was to teach astronauts variations of the AGSM. Apollo 11 pilot, Michael Collins, recalls his fuge experience in his 1974 autobiography *Carrying the Fire*: "If you breathe normally, you find you can exhale just fine, but when you try to inhale, it's impossible to re-inflate your lungs, just as if steel bands were tightly encircling your chest."

Another important project being conducted at Johnsville was testing body support couches. A group of National Advisory Committee for Aeronautics (NACA) aerodynamicists had designed an extremely strong and lightweight couch made of fiberglass, which could be contoured to custom-fit the body dimensions of an astronaut. In 1958, the couch was tested at Johnsville to see which body position would best tolerate G. The couch had been molded to fit the anthropomorphic dimensions of Robert A. Champine, one of NACA's leading test pilots. Champine rode the Johnsville centrifuge to a peak of 12 G. The following day, Navy Lieutenant Carter C. Collins volunteered to test the couch. Since his frame was smaller than Champine's, the Johnsville scientists packed foam-rubber padding into the recesses of the fiberglass bed. Collins climbed into the gondola and prepared himself for the first run. After tolerating 12 G on the first run, Collins endured five more runs, the last of which peaked at 18 G. Then, on the sixth run, Collins withstood a peak of 20.7 G, applied transversely for six seconds. Later that day, R. Flanagan Gray, a physician at the Johnsville laboratory, rode the centrifuge with the contour couch and also endured a 20-G peak. The acceleration patterns which Collins and Gray were exposed to corresponded to a re-entry angle of 7.5°. The NACA engineers, already working overtime on designs for a manned orbital capsule, were elated, since it seemed they finally had an effective anti-G device that was small enough and light enough to fit into the space capsule that was to be used for the initial manned space mission. But the custom couch was only one component that was ultimately used to protect astronauts. To provide the Mercury astronauts with as much protection as possible, the engineers combined the use of restraining straps, a semi-supine posture, frontward application of acceleration loads, and the reversal of the spacecraft attitude during orbit to permit frontward imposition of re-entry loads as well.

While the engineers were experimenting with various permutations of straps, posture, and spacecraft attitudes, researchers continued to investigate other ways to protect the astronauts from G. One of these projects involved the use of the Iron Maiden. The Iron Maiden project began when one inquisitive scientist spun a fish and discovered the fish tolerated high Gs without any apparent side effects. After witnessing the spinning fish test, someone had the bright idea (or at least it seemed bright to the scientist suggesting it) of spinning a human encased in water! The theory was that the water would dissipate the G-forces, allowing the subject to tolerate high G. To the G scientists, it sounded like a reasonable hypothesis, so Gray designed an aluminum capsule roughly in the shape of a seated human that could be filled with water. The contraption was christened the Iron Maiden. Gray stayed alert throughout a 25-second run-up to 32 G, suffering only mild sinus pain. He wanted to go to the full 40-G capability of the centrifuge, but the Maiden was too big to fit inside the gondola and had to be mounted farther inward along the arm, where 32 G was the maximum acceleration possible. Nevertheless, Gray's unorthodox Iron Maiden ride established a new record for tolerance to centrifuge G-loads.

The Iron Maiden wasn't the only cutting-edge and slightly crazy centrifuge idea. In the late 1950s, two scientists, Carl Clark and Dr James Hardy, Johnsville's Research Director, calculated that if a spacecraft could be steadily accelerated at 2 G, it could reach the Moon or Mars in days or even hours. The problem was no one knew whether a human could survive the constant acceleration. Clark, who was the Centrifuge Training Officer for the X-15 pilots at Johnsville at the time, decided to use the centrifuge to find out. With the approval of Dr Hardy and Navy management, Clark had his reclining chair and a small electric stove installed in the centrifuge gondola and, on the weekend after Thanksgiving in 1957, became the first and only human to experience 2 G for 24 hr. During the run, Clark was able to cook and eat, and converse cogently with the medical staff. He also slept for a number of hours but, as time dragged on, listening to music became the activity of choice. When the 24 hr had elapsed, he climbed out with only a little unsteadiness and then returned home for a good dinner and a night's sleep. Additional marathon rides were planned, but these were discouraged, partly because of the mounting evidence of the effects of G.

One factor that concerned the researchers back in those days was the insult to the brain due to the lack of oxygen in the blood. Then there were the effects of GLOC, motion sickness, disorientation, anxiety, euphoria, and confusion. Other effects encountered by centrifuge riders included swelling of the feet and ankles, ruptured blood vessels in the groin area, blood clots, and temporary changes in blood-flow patterns in the lungs. Scientists also warned of the possibility of collapsed lungs, fractured ribs, and chest pain. While many of these effects were transient and fairly rare, they weren't dismissed, so potential centrifuge riders had to submit to extensive medical testing. Once cleared, subjects generally rode the centrifuge in one of two modes: either closed-loop or "dynamic flight simulation", in which the rider had full control over the movements of the centrifuge; and open-loop, or "meat in a seat", with the rider basically being used as a lab rat. During their runs, riders were given various tasks to perform, such as flying simulated combat missions, all the while being monitored by doctors who were able to stop the ride if anything happened.

Despite the discomfort and dangers, researchers never had a problem recruiting volunteers. Many were simply interested in seeing what their G-limits were, others thought the experience would make a good story, and some simply wanted an excuse to get out of the office! For their efforts, subjects were given a souvenir – a videotape of their ride to show friends and family how they looked when they lost consciousness (we still do this today).

In addition to solving the problems of protecting astronauts during launch and re-entry, the centrifuge research conducted at Johnsville and other centrifuges has had a lasting impact on the training of military pilots, the development of anti-G suits and techniques, and the design of aircraft and spacecraft systems. The Johnsville researchers also studied practical problems, including the disorientation of Navy pilots following night catapult launches from a carrier and spin recovery techniques in fighter aircraft such as the F-14 Tomcat. The last decade of operations at Johnsville saw one of the centrifuge's most important contributions: the Gender Neutral Study (GNS). The GNS came about in the 1990s as a result of a mandate from Congress that women and men should be able to fly fighter jets. One of the obvious GNS questions that needed to be answered was whether females could hold their own in the high-G environment. After testing in the Johnsville centrifuge, it turned out that women could more than hold their own against the flyboys. The tests confirmed that women have comparable acceleration tolerance with the men and, by all accounts, they didn't complain nearly as much as the male subjects did.

In 1996, due to mounting costs, the Johnsville centrifuge was forced into retirement. As we'll see, while centrifuge work continues at other US military and NASA centers, the center of the action seems to be shifting overseas, with new centrifuges in Sweden, China, and Japan, although none measures up to Johnsville's capabilities. When it was decommissioned, the centrifuge was disassembled and transferred to the Smithsonian. The gondola was later sold by the Smithsonian and transferred back to the Johnsville Centrifuge and Science Museum. Of all the world's centrifuges, the Johnsville G Machine played perhaps the most important role in shaping the future of space travel, since, for nearly 50 years, literally *every* American astronaut took a ride in it.

Holloman

While Johnsville carved a niche in trouble-shooting the effects of launch and re-entry Gs, Holloman Air Force Base (HAFB), New Mexico, was the epicenter of *pilot-related* G research and training. Since the centrifuge (Figure 4.10) was first certified at Holloman in 1988, it was the only USAF-owned human centrifuge used for this training. The Holloman centrifuge, which was configured for anything from the T-38 to the F-22 and every high-G aircraft in between, was used to train more than 30,000 students from all military branches, North Atlantic Treaty Organization (NATO) troops, and members of allied forces. Not surprisingly, when it was decommissioned in 2010, it was the busiest, most utilized centrifuge in the world, having served not only to train pilots, but also to conduct all sorts of G-related studies and projects.

While the last centrifuge mission at Holloman was bittersweet for many of the physiology technicians who had spent years teaching in the classroom and working with the centrifuge, their disappointment was tempered somewhat by the knowledge that a new centrifuge was being built at Wright-Patterson Air

Figure 4.10 The USAF centrifuge at Holloman Air Force Base is operated by the aerospace physiology department for the purpose of training and evaluating prospective fighter pilots for high-G flight in Air Force fighter aircraft. Image courtesy: USAF.

Force Base. The new $34.4 million centrifuge, built by Environmental Tectonics Corp., of Southampton, Pennsylvania, will spin at the base's new Human Performance Wing complex starting in 2012. Once certified, the human performance gurus who worked at HAFB will continue their work.

NASTAR

When NASA opened for business on October 1st, 1958, its most important job was to send an American into space and return him safely to Earth. To achieve this goal, Project Mercury would use one of several dependable military rockets to launch a small, one-man spacecraft that would return to Earth and land in the ocean. Astronauts would be chosen for the program from the best military test pilots who had education in science or engineering. It was a simple idea, but making it happen was anything but. Thousands of scientists, engineers, technicians, and other workers were needed and thousands of millions of dollars were called for to fund the enterprise. But Congress approved the money and NASA organized the program. The McDonnell Company designed and built the spacecraft while the Army and Air Force built the rockets. NASA announced the seven astronauts it had chosen on April 19th, 1959, and they wasted no time in beginning training. Test flights began later that year, although these test flights didn't carry astronauts. A series of successful (and less-than-successful) flights eventually paved the way for NASA's first astronaut. On May 5th, 1961, Navy pilot, Alan Shepard, squeezed into his tiny Mercury spacecraft and waited. And waited. First, the weather caused a delay and then last-minute repairs needed to be made to his radio system. Tired of waiting, he asked the ground crew "Why don't you fellows solve your little problems and light this candle?" Finally, they launched the rocket. Shepard's suborbital flight lasted only a few seconds longer than 15 minutes, but he flew 187 km high and 480 km from the launch pad. The first manned flight of project Mercury was a complete success.

Thanks to the new generation of civilian space adventurers, "light this candle" is a phrase that has been re-launched in spaceflight's lexicon but, while Alan Shepard was ensconced atop a Mercury Redstone rocket when he spoke those words, the new breed of space explorer can get his or her candle lit on the ground in a suburb of Philadelphia long before heading into orbit. As part of the National Aerospace Training Center's (NASTAR) space launch training program, aspiring astronauts willing to pay $6,000 can discover how it will feel to be launched into orbit and whether they have the right stuff to cope with the rigors of suborbital flight – a sensible idea before forking out $200,000 for Virgin Galactic's (Figure 4.11) suborbital ticket price.

For their $6,000, budding suborbital astronauts also get a flight in NASTAR's altitude chamber as part of their high-altitude indoctrination training and receive instruction on the subject of aerospace physiology. The highlight of the course, however, is riding in the company's state-of-the-art space simulator – the STS-400 Phoenix (Figure 4.12).

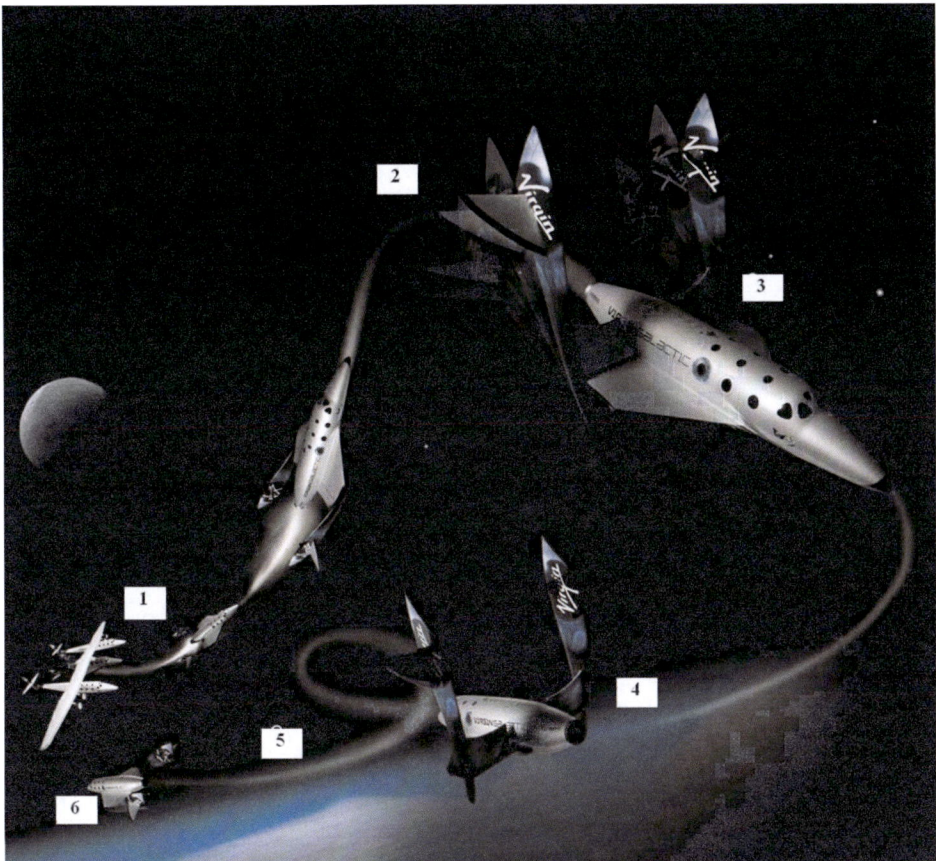

Figure 4.11 The flight profile of Virgin Galactic's SpaceShipTwo. Image courtesy: Virgin Galactic/Scaled Composites.

The STS-400 (the STS acronym stands for Space Training Simulator) is a space module that sits at the end of the swing arm (also referred to as the "planetary arm") of NASTAR's AFTS (the AFTS acronym stands for Authentic Tactical Fighting System)-400 centrifuge (Table 4.1). The module comprises a single-seat cockpit representative of a generic seat in a suborbital space vehicle. The STS-400 can provide simultaneous Gz-forces and Gx-forces that simulate the rocket ignition and pitch-up maneuver of a suborbital rocket (such as SpaceShipTwo) after being released from a captive-carry aircraft (such as WhiteKnightTwo). The STS-400–ATFS combination is the most advanced motion-based flight simulator in existence.

Thanks to its pitch and roll capabilities, the STS-400/ATFS-400 can model the flight profile of just about any aircraft in existence. This capability, combined with authentic modeling of the cockpit and immersive real-world visual displays, delivers the most authentic flying experience short of actually flying the aircraft or spacecraft.

Figure 4.12 NASTAR's Space Training Simulator (STS-400) creates an authentic spaceflight experience complete with satellite imagery visuals and dynamic G-forces on launch and re-entry. The STS-400 uses the same flight-training motion-based technology of the ATFS-400 "Phoenix" to accurately flight test and fly suborbital and orbital spacecraft missions. With interchangeable cockpit and passenger seat modules, the STS-400 is an ideal platform for both pilots and passengers to master the unique challenges of the space launch environment. Image courtesy: NASTAR.

Table 4.1. Parameters and characteristics of the AFTS-400/STS-400.

Parameter	Characteristics
Gz range (along spine)	−8 to 12 Gz
Gx range (through chest)	±8 Gx
Gy range (side-to-side)	±6 Gy
Max mean G-onset	8 G/sec from 1.4-G idle
Pitch motion range (forward and backward tilting)	±360° continuous
Roll motion range (side-to-side tilting)	±360° continuous
Main drive motor	1,250 HP DC motor
Maximum gondola pitch axis angular speed and acceleration	100°/sec and 250°/sec
Maximum gondola roll axis angular speed acceleration	100°/sec and 200°/sec

Figure 4.13 Greg Olsen is an American entrepreneur, engineer, and scientist who, in October 2005, became the third private citizen to make a self-funded trip to the International Space Station. Image courtesy: NASA.

NASTAR's space-training program was born out of necessity for one of the original citizen spaceflight participants – Greg Olsen. Olsen (Figure 4.13), a New Jersey businessman, had tried to secure a $30 million seat on a Russian Soyuz launcher for a 10-day visit to the International Space Station (ISS) through Virginia-based Space Adventures. After washing out of the qualification process in Russia in 2004, he researched manufacturers of human-rated centrifuges and ended up at NASTAR. He wasted no time contracting them to create a Soyuz launch and re-entry profile on the centrifuge. Once that had been done, Olsen started G-tolerance training and was accepted the second time around, becoming the third spaceflight participant to visit the ISS in October 2005.

Two years later, spurred by the advent of commercial suborbital flight ventures, spawned in part by the flights of SpaceShipOne (Figure 4.14), NASTAR

Figure 4.14 SpaceShipOne was a high-altitude research rocket, designed for suborbital flights to 100-km altitude. The unique configuration allowed aircraft-like qualities for boost, glide, and landing. The ship converts (via pneumatic-actuated feather) to a stable, high-drag shape for atmospheric entry. Image courtesy: Virgin Galactic/Scaled Composites.

officially launched its space launch training program with Sir Richard Branson's Virgin Galactic as a client. Branson initially contracted NASTAR for the first group of 100 Virgin Galactic reservation holders to take the training to prove his business model, which made sense, since the company had to know whether the general population could tolerate suborbital space training.

At the time of writing, the guidelines for spaceflight providers are listed in FAA Part 460 [1]. This regulation states that suborbital space training should include mission-specific flight crew training for each phase of flight using some type of simulator as well as demonstrating that the crew have the "ability to withstand the stresses of spaceflight, which may include high acceleration or deceleration, microgravity and vibration, in sufficient condition to safely carry out his or her duties so that the vehicle will not harm the public". That's for the flight crew. The passengers? Well, the only guideline for passengers at the moment is that they provide written informed consent confirming that they understand the risks. As part of its contract with Virgin Galactic, NASTAR required the company's astronaut wannabes, who ranged in age from 18 to 88, to complete a questionnaire that was reviewed by physicians. Once the medical review was out of the way, the Virgin Galactic trainees spent two days learning about space physiology in the classroom and were taught the skills to help them tolerate Gs in the centrifuge. Since the first course in 2007, NASTAR has received a safety approval from the FAA for its spaceflight training programs.

While NASTAR trains suborbital tourists and scientists, its biggest customer is the Air Force, but just how closely does an Earth-bound flight simulator mimic the force and motion cues of a fighter jet? Can the gimbaled AFTS-400, with its pitch and roll capability, *really* create the illusion of flight? Well, the gondola/cockpit is fitted with a production hardware control stick, rudder pedals, and throttles as well as a partial dome color visual system that projects external scenery in the pilot's forward field of view (FoV). It's a convincing replication of the actual aircraft. Before answering the question of whether it can replicate flight though, it's helpful to understand the concept of transfer of training, since this is the criterion by which flight simulators are measured.

Simulation

In the world of aviation, transfer of training refers to the measurable and beneficial effects of learning or practicing a skill or task in a simulator that can be applied directly to the aircraft. For transfer of training to occur, it is obviously necessary that the simulator and the aircraft share common elements – the better the elements of each correspond with the other, the higher the transfer of training, and vice versa. But simply maximizing realism by means of high-fidelity representations in the AFTS-400 cockpit simulator isn't enough to maximize transfer of training. To do that, the simulator also needs to provide sensory cues (Table 4.2). What do we mean by this? Well, pilots interact with their external environment while performing mission tasks. They perceive aircraft state changes, maintain external and internal cockpit awareness, make decisions, modify aircraft control strategies, and make all sorts of mission-specific tasks. Now, while emulating the man–machine interface is relatively straightforward, the simulation of the in-flight environment is another matter. How do you mimic the aircraft's maneuvering dynamics, external cues, *and* the gravity accelerations and decelerations to a fidelity that tricks the rider into thinking they are actually in a real aircraft? It's a tough ask.

In an attempt to provide pilots with the highest-fidelity experience possible, large commercial aircraft simulators employ force and motion cueing through the use of 6-DoF of freedom Stewart platforms. In these, the force and motion cues are specifically tuned to the performance capabilities of commercial aircraft and include cues such as runway rumble, buffeting, and retraction of landing gear. For the most part, the simulators do a good job providing commercial aircrew with a high-fidelity flying experience, but tactical fighter aircraft have to operate in a much larger flight envelope. As we've seen in Chapter 2, pilots have to deal with rapid-onset, high-magnitude accelerations, which provide a

Table 4.2. Sensory cue categories.

Cue category	Description
Visual	Changes in external scene such as weather and sensor and display symbols
Gravito-inertial acceleration	Positive and negative Gs, roll, pitch, yaw, inner-ear sensations
Physical airframe	Vibration, turbulence, buffeting
Aural	Cockpit noise, engine noise
Environmental	Airflow, oxygen flow, temperature

Figure 4.15 The F/A-18E/F Super Hornet is a twin-engine carrier-based multirole fighter aircraft capable of pulling more than 7 G. Image courtesy: USAF.

constellation of contrasting sensory cues. For example, the flight control system of the Super Hornet is capable of commanding maximum Gz accelerations of +7.5 and –3 G for symmetrical maneuvers and +6 and –1 G for unsymmetrical maneuvers.

Then there are the roll rate capabilities. We'll use the Super Hornet (Figure 4.15) again as an example. This aircraft can roll at 200° per second, which is an important consideration when determining the suitability of a centrifuge-based simulator, many of which can't be tuned to provide the same magnitude of roll rates. Another consideration for centrifuge-based simulator designers is the emulation of departures from controlled flight and out-of-control flight (OCF). You see, fighter pilots are taught to use tactile sideforce cues to recognize departure from controlled flight and when recovery has begun. This sideforce is felt in the cockpit as a sideways push and is a reliable feedback indicator to the pilot that departure from controlled flight continues. Once departure from controlled flight has occurred, the pilot has to recognize more cues caused by uncontrollable changes in angle of attack and airspeed.

But departure from controlled flight and OCF are just two types of maneuvering the fighter pilot may experience. What about all the mission subsets and mission-representative tasks? These subsets, some of which are listed in Table 4.3, comprise a diverse set of embedded tasks such as maneuvering the aircraft, cockpit communication, and operating the aircraft and weapon systems.

As you can see from the lists of subsets, the cockpit is a busy place, so how do centrifuge simulator designers manage to emulate all the complex tasks facing the pilot? Well, one way is to analyze aircraft data files from operational fighter squadrons and evaluate them by comparing what the AFTS-400 can do, which is what the US Navy did in 2008. By performing this evaluation, the scientists analyzed mission subset and acceleration spectrum data for each mission and identified the most important factors that could be expected from a centrifuge-based simulator. Obviously, the centrifuge doesn't have a problem emulating the G-forces, but what about the *fidelity* of those lateral/sideforce accelerations

Table 4.3. Fighter mission subsets.

Mission area	Mission subsets
Air-to-air warfare	• Beyond-visual-range engagements
	• Within-visual-range engagements
Air-to-ground warfare	• Precision weapons delivery
	• Conventional weapons delivery
	• Surface-to-air counter-tactics
Basic proficiency	• Formation flight
	• In-flight refueling
	• Emergency situations/crew resource management

mentioned earlier? Well, studies that have investigated the sensory expectations of pilots have revealed that the majority of aircrew found the centrifuge simulator to be very good at emulating the training environment. The pilots found the large-magnitude accelerations were similar to the fighter aircraft, the only difference being that the AFTS-400 lacked the excess power of the Super Hornet. The pilots were also impressed by the AFTS-400's ability to roll and pitch at angular rates comparable to the Super Hornet. Thanks to this roll-and-pitch capability, the AFTS-400 is also able to distribute the acceleration vector along the three gondola body axes, which means that those sideforce accelerations mentioned earlier can be experienced.

There were a few minor drawbacks, however.

One of the shortcomings was the fact the AFTS-400 only rotates in one direction (clockwise when looking along the rotational axis) and it can't reverse its direction of turn, which means the centrifuge control system has to provide well-tuned manipulation of the gondola pitch and roll angles to ensure the pilot perceives the difference between left and right turns. Given the versatility of the AFTS-400, this minor drawback wasn't deemed a serious problem. What did cause pilots some consternation was the overwhelming tumbling sensation caused by head movement and also by the Gz offset. The latter sensation is caused partly by the angular deceleration of the arm of the centrifuge and partly by the gondola swinging down towards its vertical stationary position. This displaces the head with reference to the plane of rotation and it is this change in head position that increases stimulation in one pair of the semicircular canal receptors whilst decreasing the stimulation in another pair. As you can imagine, it would be difficult to perform tasks inside the cockpit when experiencing such a disorienting sensation, which is exactly what the pilots reported. Many were affected to the extent that they could only focus on one cockpit task while the tumbling occurred.

So, does a centrifuge-based simulator compromise training effectiveness and, if so, can anything be done to make it more realistic? Well, first let's once again underline the difference between what a fighter jet can do and what the centrifuge can do. Consider the following maneuver: an F-18 aircraft at 420 knots at 5,000 m has sufficient power to transition within one second from non-maneuvering flight at 1 G to maneuvering flight at 5 G while maintaining constant airspeed and generating a resultant turn radius of approximately 1,000 m. The centrifuge, constrained to an 8-m swing radius, must, within the same second, nearly double its tangential speed from approximately 10 m/second to almost 20 m/second while simultaneously coordinating a bank-angle change of approximately $34°$ to correctly resolve the resultant acceleration along the new Gz axis. It's a tough challenge for a centrifuge to emulate because even state-of-the-art centrifuges like the AFTS-400 only have 3 DoF available but must somehow generate the linear and angular acceleration cues for an aircraft that has 6 DoF. In short, such emulation is impossible, but there are some strategies that can be incorporated to help centrifuges match more closely the aircraft's rectilinear accelerations. For example, the AFTS-400 employs a G-pointing motion-control system that continuously points the gondola relative to the acceleration vector so the three components of desired linear acceleration in the "aircraft-representative" coordinate frame are achieved. The resultant acceleration vector, which is produced by the combination of normal and tangential accelerations of the swing arm, is distributed along the three components of the "aircraft-representative" frame of reference by selecting appropriate gondola pitch and roll angles. It's a versatile simulation that is only marred by the pilot's sensitivity to the angular acceleration in the three axes, which results in some disorientation. To counter this, the G-pointing strategy incorporates washout algorithms to coordinate the combination of centrifugal and angular acceleration at levels below the threshold of detection.

While the G-pointing strategy and the washout algorithms solve many of the disorientation problems, one issue that can't be solved is the problem of head motion while under G. When a pilot moves their head out of alignment with the gondola, the angular acceleration can't be corrected by washout algorithms or any other strategy. It's the reason I tell pilots to keep their gaze steady on the computer-generated aircraft during their G runs – just one glance away from the screen is enough to provoke disorientation and unpleasant nausea-like sensations that sometimes result in motion sickness (which is why every pilot that rides our centrifuge is given a "boarding pass" – just in case). While staring at the X -

Figure 4.16 Star City centrifuge. Image courtesy: Wikimedia Commons.

Figure 4.17 Star City centrifuge gondola. Image courtesy: Wikimedia Commons.

Box chase plane may solve the problem of vestibular sensations in the centrifuge, it's not a very helpful strategy for a pilot flying an aircraft because they have to constantly make head movements, scanning outside and inside the cockpit for visual information. Any symptoms of motion sickness will obviously distract and disrupt the pilot's normal cockpit routine and decrease the effectiveness of training, which is why even the best centrifuge simulators can't be used for full mission training scenarios.

While the ATFS-400 may not be able to simulate the full spectrum of mission training, it offers unique training capabilities for users, both research and training alike. In addition to being able to emulate the performance characteristics and physiological stresses of the most advanced fourth- and fifth-generation fighter, NASTAR's fuge is an affordable, high-fidelity, G flight simulator that represents the current state-of-the-art high-performance "flyable" motion system.

Star City

Finally, no chapter on the subject of centrifuges would be complete without mentioning the TsF-18. Located at Star City, the TsF-18 is the largest centrifuge in the world. Star City is an hour's drive north-east of Moscow. The cosmonauts' home, which looks more like a housing development thanks to the trees lining the roads, is a very different setting from NASA's Johnson Space Center (JSC) in Houston, but the military sentry standing duty reminds visitors that this is no ordinary suburb. Among the myriad space-training hardware is the 20-year-old Hydrolab, a gigantic swimming pool equipped with a submersible platform that can hold up to 7 tonnes' worth of space station mockups and, in a nearby building, cosmonauts go through their annual spins on the TsF-18 (Figure 4.16) – they have to pass a 40-second chest-to-spine 4 G and a 40-second chest-to-spine direction 8 G. The rest of the world's centrifuges are nothing compared with the Star City behemoth. Weighing 300 tonnes, with an 18-m arm capable of spinning up to 30 G, the TsF-18 can house up to 350 kg in its capsule (Figure 4.17), which is fixed in a gimbal suspension and has three degrees of freedom, which allows the G-load vector to be oriented in any direction.

Reference

[1] www.faa.gov/about/office_org/headquarters_offices/ast/regulations/, Part 460 – Human Space Flight Requirements.

5

Formula One

The monocoque chassis is barely big enough to accept the driver, the fuel tank, the radiators, and the myriad electronic systems that control the car. Carbon-fiber suspension arms are directly mounted to the monocoque via titanium fixtures but there are no joints, since suspension movement is accomplished by flexure. At the front, there are upper and lower control arms, which actuate pushrods that work tiny shocks via bell cranks and torsion bars that are used for springing. At the back, the suspension mounts directly onto the gearbox and there are upper and lower control arms, with pushrods acting on the shocks and torsion bars via bell cranks. The brakes are massive carbon-carbon discs and the calipers have six pistons.

G in Formula One

The Formula One (F1) car (Figure 5.1) is an exquisite piece of machinery capable of jaw-dropping, mind-numbing speed thanks to a V-8 engine that displaces 2.4 liters, runs up to 19,000 rpm, and generates a power-to-weight ratio of 4.5 pounds per horsepower (by comparison, the $1.7 million Bugatti Veyron, the quickest and most powerful street-legal car ever built, pushes out a puny 1.8 pounds per horsepower!). Being able to get the most out of an F1 car requires phenomenal reflexes, extraordinary hand-to-eye coordination, outstanding depth perception, and a superhuman ability to analyze and assess how the car is performing at any given instant, because these cars are fast.

Real fast.

An F1 car accelerates from 0 to 100 km/hr in 2.8 seconds, which is 0.3 seconds slower than the Bugatti Veyron, but the race car reaches 160 km/hr from a standstill in 4.1 seconds, which is 1.2 seconds quicker than the Bugatti (an F1 car can do a quarter-mile in 9 seconds incidentally). From the perspective of the on-board cameras, the likes of 2009 World Champion Jenson Button make driving an F1 car look effortless. It's anything but. F1 drivers are probably the fittest all-round athletes on the planet and they have to be because the driver of one of these awesome cars is subject to massive vibrations, extreme heat, and an intense amount of physical work over one and a half hours of racing. And then there's the acceleration.

The G-forces[1] produced by an F1 car are simply unbelievable. Thanks to the

---

[1] The accelerometers that are installed in every F1 car – Model 4203 Triaxial Accelerometer built by Measurement Specialties – were originally designed for missile systems and record a tri-axial G-force measurement 20 times per second.

Figure 5.1 Formula One (F1) cars are capable of going from 0 to 160 km/hr and back to 0 in less than five seconds. During a demonstration at the Silverstone circuit in Britain, an F1 McLaren–Mercedes car driven by David Coulthard gave a pair of Mercedes–Benz street cars a head start of 70 seconds, and beat the cars to the finish line from a standing start, over a distance of just 5.2 km. Typical acceleration times include the following: 0 to 100 km/hr: 1.7 seconds; 0 to 200 km/hr: 3.8 seconds; 0 to 300 km/h: 8.6 seconds. Accelerating to 200 km/hr in less than four seconds subjects the driver to 1.45 G (14.2 m/s^2). As well as being fast in a straight line, F1 cars also have incredible cornering ability, which produces some neck-punishing lateral Gs. Image courtesy: Wikimedia Commons.

almost extraterrestrial levels of downforce and grip, the cornering speeds are so high that an F1 car feels like it's on rails. These exotic racing vehicles regularly produce a force of 4 G through corners, which means a driver who weighs 70 kg instantly weighs 280 kg. It's a phenomenal strain on the body and in particular on the parts of the body that have to move, such as the head and the arms (the effects on the torso are less of a problem thanks to the support of the seat belts). Take Turn 1, Abbey – one of the fastest corners on the Silverstone Grand Prix circuit. Abbey is a flat-out right-hander after the new pit complex that is taken at approximately 290 km/hr. When negotiating the corner, drivers experience a peak lateral force of 4.8 G (over 4 G for 1.3 seconds), while the car experiences a peak vertical force, including car mass, of 22 kN – equivalent to 2.2 tonnes. This means the car generates two and a half times its weight in downforce in the corner. Worse is to follow. The Becketts complex (Figure 5.2) includes five corners, through which the drivers experience extremely high G-forces in opposite directions within an extremely short space of time. To negotiate Becketts, drivers need a perfectly balanced car, agile changes of direction, and speed. Lots of speed. Thanks to the downforce, the drivers barely touch the brakes when negotiating the complex set of corners, which means the G-forces pile up. The sequence of G-forces is as follows:

Turn 10 (Left, 300 km/hr) – **2.2 G**
Turn 11 (Maggots, Right, 275 km/hr) – **4.8 G**
Turn 12 (Becketts One, Left, 230 km/hr) – **3.9 G**
Turn 13 (Becketts Two, Right, 195 km/hr) – **3.9 G**
Turn 14 (Chapel, Left, 240 km/hr) – **2.2 G**

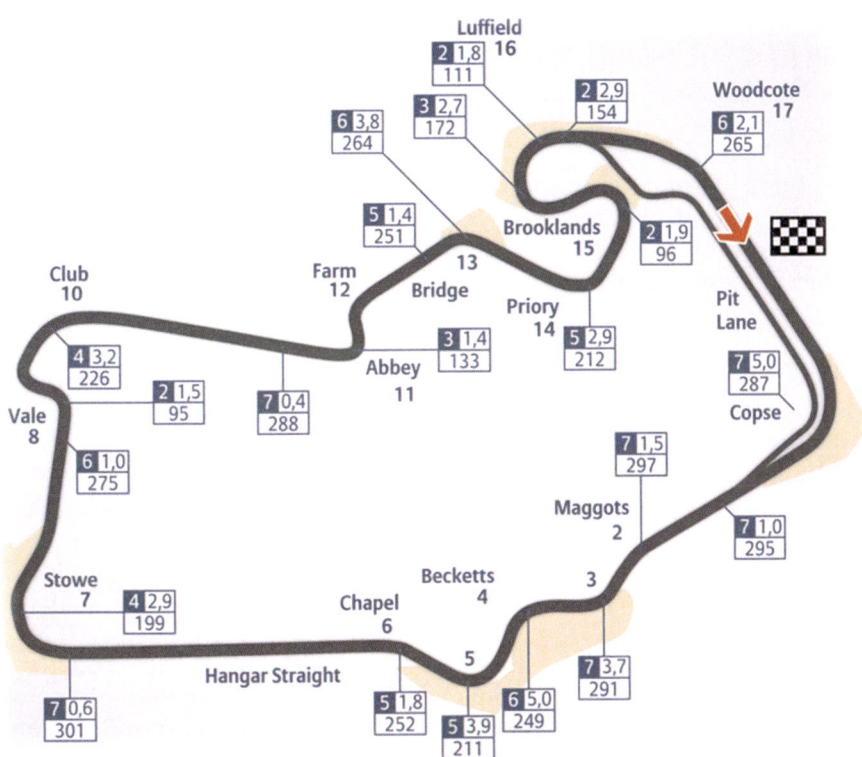

Figure 5.2 Most Formula One drivers agree that the Maggots–Becketts–Chapel complex of Silverstone is superb. With the exception of Eau Rouge, the complex is probably the most G-demanding and technical set of turns of any Grand Prix circuit in the world. Image courtesy: www.formula1journal.com.

Imagine driving through Maggots. Your head would weigh almost 20 kg! Of course, you can't simulate this in a road car – well, not unless you strap a few bricks to your head and shake it from side to side for 90 minutes. The same acceleration forces are exerted on the driver's arms as they work the steering wheel. Again, there's just no way you can simulate these sorts of forces in a road car, unless you take those same bricks and strap them to your arms next time you go for a drive.

But F1 racing is much more than just cornering. To negotiate those corners, F1 drivers have to brake and, with 16 or 17 corners on each lap and 60 or 70 laps in a race, that's an awful lot of braking. From a top speed of around 320 km/hr, applying full braking force in an F1 car will produce 6 G, which is just 2 G short of the force required to render a human unconscious. The deceleration is so extreme that the braking forces literally suck the water out of the driver's tear ducts. I haven't driven an F1 car (I wish I could say I had though), but I have ridden in a centrifuge as part of my job as Director of Canada's Human Centrifuge Operations. I've also ridden in the back seat of a Hawk, a nimble jet capable of pulling 6 G in the blink of the proverbial eye. My jet ride lasted just over an hour, during which the pilot put us through several high-G maneuvers; 2-G and 3-G turns weren't so bad but, once the Gs ratcheted up to 5 and 6, my breathing became labored and I was grateful for my G-suit. I can't imagine enduring those sorts of G-forces for 90 minutes straight. It's one hell of a workout.

Another type of acceleration the F1 driver has to contend with is vibration. Massive vibration. It's because F1 cars don't have conventional suspension. Unlike road cars, the comfort of the driver doesn't enter into the equation when it comes to designing a suspension system. This means the spring and damper rates are extremely firm to ensure that the impact of hitting any bumps and curbs is dissipated as quickly as possible. Perhaps the best analogy is to think of catching a ball rather than letting it bounce – the springs absorb the energy of the bump and the shock absorber releases it on the return stroke and prevents an oscillating force from being generated.

This stiffly sprung suspension makes for some rough rides, since even the smallest bump results in a huge jolt inside the cockpit, so the driver feels as if he (there are no female F1 drivers) is being shaken apart. The cars are loaded with sensors so engineers can measure the magnitude of the vibration accelerations. The maximum vertical vibration acceleration experienced (suffered!) by the drivers is 3 G. These vibrations are random and are only experienced for split seconds, but they occur at just the wrong frequency that humans find uncomfortable. Unlike many of the challenges of driving an F1 car, no amount of training can accustom a driver to the effects of these vibrations. The short and long-term side effects are not inconsiderable. During a race, the driver will experience elevated heart rate, increased muscle tension, and slightly impaired vision but, in the long term, drivers may suffer from spine and nervous system problems, which is probably why arthritis is an occupational hazard in this elite group of drivers.

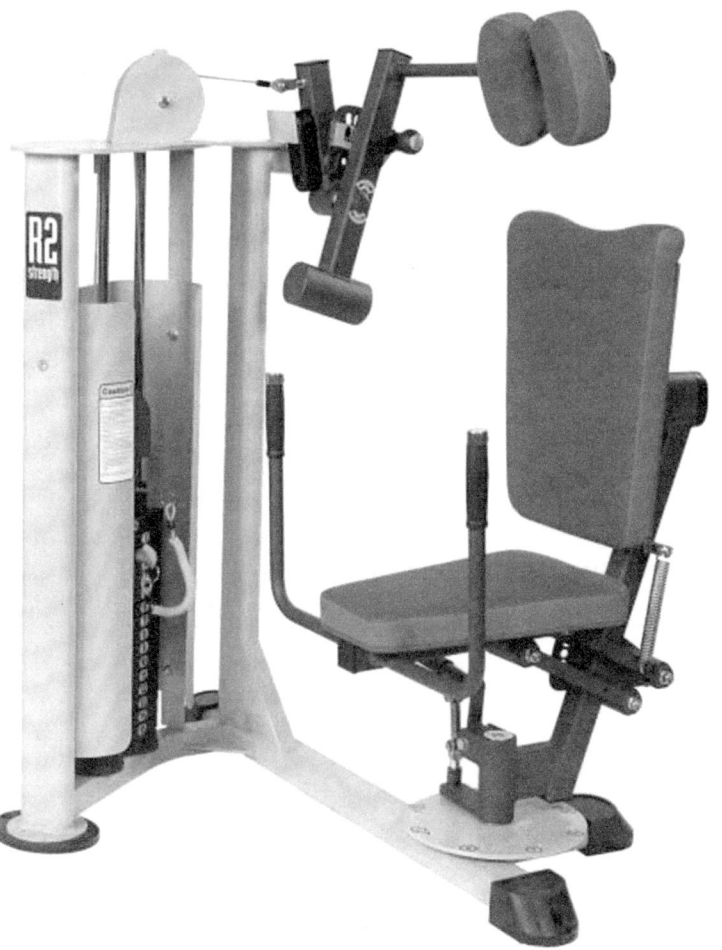

Figure 5.3 There are few sports that place as many demands on neck muscles as Formula One (F1). Common injuries incurred by F1 drivers tend not to be breaks or dislocations, but bruises and strains inflicted by high-G turns, cornering, braking, and accelerating. A human head weighs around 6 kg and a helmet weighs in excess of 1 kg. During a typical high-G F1 corner, a head will weigh more than 40 kg! Image courtesy: Leisure Lines.

"There is no other sport as physically demanding as Formula One." So says Renault driver, Heikki Kovalainen. I'd tend to agree with him, despite the fact that I've raced ironman triathlons and 100-km running races. Given the tremendous stresses imposed on the F1 driver, it's not surprising that drivers spend an awful lot of time training and the anatomy of an F1 driver is as precisely engineered as the tool of his trade. And perhaps one of the most important aspects of the driver's training regime is beefing up the neck and back muscles, which is why all drivers develop outsized necks by using neck trainers (Figure 5.3).

Impact Gs

The first thing you saw was flying earth as Robert Kubica's BMW-Sauber chewed through the grass, bouncing heavily out of control, the car far off the racing line within the shadow of the hairpin and its hard-braking zone. What Kubica probably saw first was a wall coming quickly towards him.

Much too quickly.

The 22-year-old Pole appeared to have made contact with the rear of Jarno Trulli's Toyota on Lap 27 of the 2007 Montreal Grand Prix. Travelling at about 300 km/hr, he shot across the track coming out of Turn 9, through the grass, and, with sickening, car-disintegrating force, slammed into a retaining wall in the infield. Had Kubica's car been even slightly more airborne, it likely would have cleared the wall and sailed directly into the oncoming traffic that was heading out of the hairpin and towards the long, bullet-fast Casino Straight. Mercifully, the wall prevented that. Instead, still moving at sickening speed, the car reversed direction, spun, and flipped 360° across the track. A wheel flew loose of its tether and carbon-fiber shards sprayed like rain. The wreck finally came to rest on its side with a final hard slam against another wall.

Breaking down the accident, event by event, Kubica hit a total of three barriers, the first a glancing blow of his right-front that ripped the wheel off but which barely reduced the car's estimated 280-km/hr momentum. The next event was an approximately 75° frontal impact that pretty much wiped away the car's footwell and subjected Kubica to an average deceleration of 28 G. The car then rolled violently and hit the barrier on the opposite side of the circuit with much of its kinetic energy spent. As the smoking wreckage came to a halt on its side, Kubica's booted feet could be seen protruding from the shredded nose-cone.

It was an ominous sign.

The accident had fatality written all over it – uncontrolled speed meeting cement and asphalt at horrendous speed and only a carbon-fiber monocoque to prevent lethal injury. Such was the violence of Robert Kubica's Montreal accident that most onlookers initially feared the worst.

Safety crews rushed to the scene and extricated Kubica from what was left of his car before airlifting him to the trauma unit of Montreal's Sacre Coeur Hospital. Miraculously, he was diagnosed as having suffered only concussion and a sprained right ankle. Kubica's accident was by far one of the most violent the sport had ever witnessed and, while journalists described Kubica's survival as miraculous, the fact that the Pole escaped with relatively minor injuries was an eloquent tribute to the Federation Internationale de l'Automobile (FIA) and its relentless pursuit of safety.

Initially, it was speculated that Kubica took a 60-G impact but, after carefully analyzing the data from the accident data recorder (ADR),[2] it was announced that the Pole had survived a 75-G impact. Only the construction of Kubica's F1 car saved him from a potential severe injury or even death. After further analysis of the ADR data, investigators reported that all the safety features of Kubica's F1 car had complemented each other and worked to perfection, allowing the BMW-Sauber driver to survive the 75-G impact. The report issued by the investigators stated that the G-force Kubica was subjected to peaked at 75 G in a millisecond. If the force had been sustained for much longer, the Pole may have been severely injured:

> "While we were completely shocked about the violence of the accident, we were over the moon to see Robert relatively unharmed and were very content about the behavior of the chassis as survival cell."
>
> Willy Rampf, BMW-Sauber Technical Director, commenting on Robert Kubica's high-speed accident at the 2007 Montreal Grand Prix

Survival cell

The survival cell Willy Rampf mentioned was perhaps the key element in protecting Kubica from more serious injury. At the heart of each F1 car is an incredibly strong "monocoque" structure. The monocoque (Figure 5.4) is the driver's workplace and survival cell all in one. The engine is flanged onto it at the rear, the car's nose at the front. The shape of the monocoque is dictated by various factors, including the dimensions of the cockpit opening, the length of the wheelbase, the size of the fuel tank, the driver's physique, various aerodynamic requirements, and, of course, impact protection and survivability. Due to accidents such as Kubica's, impact standards have become increasingly stringent over the years, ensuring a significant increase in passive safety for the drivers. The most important tests are the frontal crash (with the nose section) at a speed of 15 m/second, a side-on impact at 10 m/second, and the stationary load test for the rollover bar, which has to withstand around 12 tonnes of pressure.

The key to the monocoque's surviving these tests lies in its design and construction. The survival cell consists of a carbon-fiber/aluminum honeycomb composite, which makes for extremely high rigidity and strength while keeping weight low. To make sure the survival cell meets the impact tests, engineers work out how many layers of carbon fiber are needed for specific areas of the monocoque. They also select specific types of carbon fiber, depending on whether forces are exerted from a single direction or several. Another strategy intended to minimize impact forces is to layer the carbon fiber. In areas subjected to particularly high loads, up to 60 layers of carbon fiber may be stacked on top

---

[2]   The ADR device measures 15 × 15 cm and has a capability of recording every piece of data for a whole race.

Test of strength
The arrows detail the loads that the monocoque must be able to withstand acc. to the FIA.

Figure 5.4 At the heart of the Formula One (F1) car is the extraordinarily tough monocoque. The ''tub'', as it is often referred to, incorporates the cockpit and the driver's survival cell, and also forms the principal component of the car's chassis. Like so much of an F1 car, most of the monocoque is constructed from carbon fiber, which makes not only for a lightweight structure, but also a very tough cell. In fact, the reason so many F1 drivers have survived extraordinary deceleration loads in accidents is due to the enormous strength of the survival cell. Image courtesy: www.flscarlet.com.

of each other. It sounds like a lot of carbon fiber, but the monocoque, often referred to as the ''tub'', is a principal part of the car's chassis, since the engine and front suspension are mounted directly onto it. This means the ''tub'' is not only a safety device, but also a key structural component, which means it has to be super, super strong.

To ensure the monocoque's robustness, high-density woven laminate panels make up its construction. The carbon-fiber weave has a coupling structure between the inner and outer layers that is bonded to the carbon fiber. The epoxy provides stability and load distribution for attachment points as well as compressive strength. Another benefit of carbon fiber is that, in tension, the elastic modulus is such that deformations are kept to a minimum, which makes for a very rigid design. The downside is that metal absorbs a huge amount of energy in the plastic deformation region whereas carbon fiber is closer to glass or elastic and will fracture.

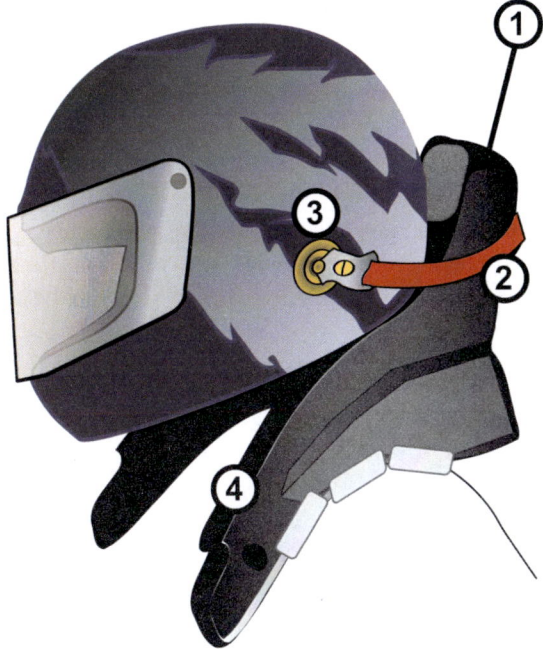

Figure 5.5 The Head and Neck Support (HANS) device is a compulsory safety item in Formula One. The purpose of the HANS is to prevent the head from whipping forward in a crash, without otherwise restricting movement of the neck. Primarily made of carbon fiber, the device is U-shaped, with the back of the U set behind the nape of the neck, and the two arms laying flat along the top of the chest over the pectoral muscles. The device is attached only to the helmet and not to the belts, driver's body, or seat, by two anchors on either side. 1: HANS device; 2: tethers; 3: helmet anchor; 4: shoulder support. Image courtesy: Wikimedia Commons.

HANS device

The slightly strange-looking piece of equipment depicted in Figure 5.5 is a Head and Neck Support (HANS for short) device, one of F1's most important safety innovations. It looks simple, but the size and shape of the HANS device and the position of the helmet anchors (located around the head's center of gravity to ensure any forces are spread evenly) are the result of several years of development.

The HANS[3] collar was originally developed in the mid-1980s by Dr Robert Hubbard, a professor of biomechanical engineering at Michigan State University.

---

[3] In 2001, in the pre-HANS era, NASCAR driver Dale Earnhardt was killed in an impact of no more than 76 km/hr because of neck injuries caused during sudden deceleration.

Figure 5.6 The HANS device is the only head and neck restraint that allows exceptional vision via slack tethers and exceptional safety due to the way it works. The slack in the tethers is automatically taken up during impact. Image courtesy: Wikimedia Commons.

The 1980s versions worked well, but they were designed for use in sports cars so they were too big to be used by F1 drivers. So, in 1997, Hubbard started working with Mercedes-Benz to develop and evaluate HANS prototypes. Gradually, with the help of Mercedes-Benz and McLaren (an F1 team that became involved with Hubbard in 1998), Hubbard refined the HANS design and reshaped it so that it would fit into an F1 car.

The HANS device, which was approved by the FIA and has been mandatory since the beginning of the 2003 season, consists of a carbon-fiber collar, two helmet tethers, and lightweight foam padding. It works by connecting a driver's head more securely to the rest of his body – a configuration that helps to control the force by which the head swings forward during the rapid deceleration of a major accident; without a HANS device, although the torso is strapped securely to the seat, the head is free to move about and, in a high-speed crash, that forward movement can be extremely violent.[4] The device fits snugly over the

---

[4] The current design of the HANS reduces head movement in a crash by 44%, the force applied to the neck by up to 86%, and the acceleration applied to the head by up to 68%.

driver's shoulders with the tall, central part directly behind the helmet. As you can see in Figure 5.6, the tethers are attached to anchors on the helmet's sides using quick-release clips and adjusted so the driver can move his head enough to drive comfortably. Once the driver is seated in the car, the seat belts are tightened so they press down on the HANS device's two "arms", thereby securing the collar in place.

In an impact, the amount of helmet movement is controlled by the tethers that you can see in the figure. It's a system that dramatically reduces the energy absorbed by the driver's head and neck, since the loading is transferred from the top of the spine to the forehead, which is better suited to taking the force. Since its introduction, the HANS device has saved dozens of drivers from serious injury and, in some cases, death. In 2004, Felipe Massa was the first F1 driver whose life was saved by the HANS device when he suffered brake failure into the very same Montreal hairpin that saw Kubica's 2007 accident (calculations showed the deceleration Massa was subject to would not have been survivable without the HANS device).

Helmet

While the HANS device does a great job in minimizing head movement, what about the helmet itself? Back in F1's early days, it was speed, not safety, that took precedence. Legendary world champion Juan Manuel Fangio preferred to race in a simple balaclava! But, by 1953, even Fangio couldn't hold back the tide of safety and helmets were made compulsory. As late as the mid-1980s, an F1 helmet weighed around 2 kg. Imagine wearing this helmet while being subjected to 5 lateral Gs. The combined weight of the helmet and a driver's head would be more than 25 kg! Not surprisingly, the 1980s-era helmet added to the risk of whiplash injuries in big shunts and also increased driver fatigue during high-G cornering and deceleration. Since head and neck trauma has been identified as the greatest single risk of injury to F1 drivers, helmet manufacturers place the greatest importance on reducing the mass of helmets, while increasing their strength and resistance, which is why today's F1 helmets (Figure 5.7) weigh just 1.25 kg thanks to their carbon-fiber and polyethylene construction. The outer shell has two layers (typically fiber-reinforced resin over carbon fiber) while the inner shell is strong plastic made of the same material as used in many bullet-proof vests.

Achieving the somewhat incompatible goal of crafting a super-lightweight helmet that is also extremely strong is a complicated process. First, the driver's head is scanned to create a life-size model. Next, this sculpted replica is wrapped, layer by layer, with 120 mats of high-performance carbon fiber to ensure a perfect fit. With every thread of fiber consisting of about 12,000 "microthreads", each of which is about 15 times thinner than a human hair, the process is as expensive as it is cutting-edge. Next, the individual layers are bonded together in an autoclave (a type of "industrial pressure cooker"), hardening under high

Figure 5.7 A Formula One helmet is extremely durable but, with a weight of about 1.25 kg, it is still relatively light and so reduces the strain on the neck and shoulder muscles of the drivers on tracks with high G-force loads. Image courtesy: Wikimedia Commons.

pressure at 132°C. Just like the carbon monocoque, F1 helmets are tested thoroughly. Fitted on an imitation metal "head" and equipped with sensors that measure deceleration, test helmets are mounted on a sled running on a vertical track. To replicate an impact, the helmet (and "head") are lifted to a specified height and dropped onto a variety of objects, including a flat surface, a sharp edge, and a half-sphere. For example, the FIA-approved helmet test facility at TRL, a research and development company based in Berkshire, UK, boasts a drop tower 15.5 m tall, which is capable of accelerating objects by up to 17 m/second (63 km/hr). A helmet designed to meet the FIA's 8,860 standards needs to withstand an impact of 9.5 m/second (34.2 km/hr). Schuberth, the F1 helmet manufacturer, demonstrates the strength of the modern F1 helmets by running over them with a 55-tonne tank!

Crash barriers

The HANS device and super-tough lightweight helmets are just two examples of safety knowledge improving, but Kubica's accident showed further advances could be made in other areas. One particular area that crash investigators focused on was the crash barriers (Figure 5.8) that Kubica almost vaulted. His car was still in mid-air as it smashed into the guard rail and, had the BMW cleared the guard rail, it would have been dead on course to hit cars coming out of the hairpin.

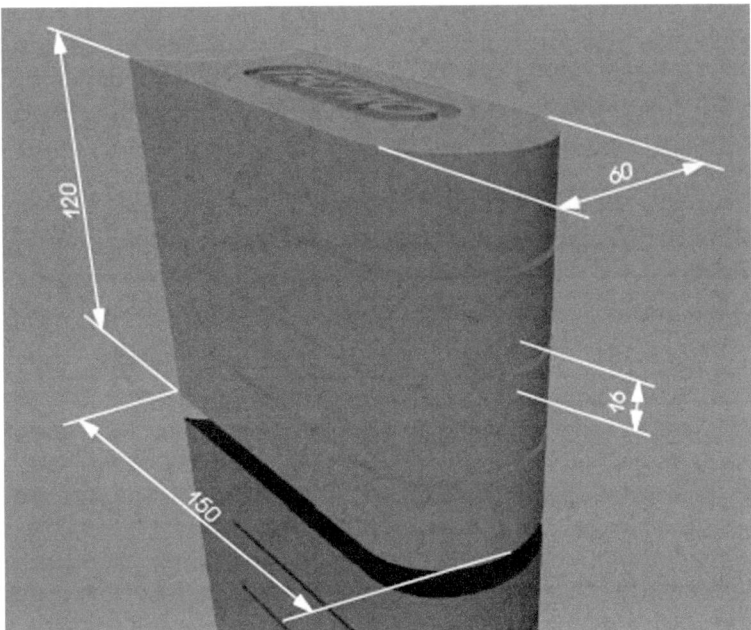

Figure 5.8 TecPro has developed a high-security barrier for use on Formula One (F1) circuits. The TecPro barriers are made from the rotational molding of flexible polyethylene while the barrier itself is filled with injected "bendy" foam of flexible density and a metallic sheet is placed at the center which reinforces the barrier. Image courtesy: TecPro.

Barriers that can protect drivers from the Gs encountered in high-speed incidents are crucial in open wheel racing. In some circuits, such as Monza and Spa, which have short run-off areas, a driver can suddenly find himself hurtling towards a barrier at speeds in excess of 200 km/hr. It's for this reason that the FIA has focused a lot of time on developing special energy-dissipating barriers designed to minimize injury and, in so doing, has revolutionized circuit safety. The genesis for this research goes back to an incident during the 1999 British Grand Prix at Silverstone when Michael Schumacher's Ferrari suffered a brake failure at Stowe corner. Schumacher hit the tire barrier at more than 100 km/hr and was, according to FIA safety expert Peter Wright, lucky to only suffer a broken leg. After Schumacher's accident, Wright contacted Hubert Gramling, a German crash-test expert, and asked him to develop a barrier that could give a driver the best chance of surviving a high-speed crash. The conversation sparked a six-year research effort in collaboration with the German automotive group DEKRA, which resulted in the design of a barrier with five separate layers designed to progressively dissipate deceleration in an impact. Initial testing didn't go well due to layers tearing but a fortuitous call to the FIA from French company TecPro International helped to get Gramling back on the right track. TecPro had designed energy-absorbing containers and wanted to help the

Figure 5.9 A Formula One monocoque being tested on a sled. Image courtesy: www.formula1.com.

FIA with its projects. The TecPro containers are blocks measuring 1.5 m long, 1 m high, and 0.6 m deep. Each end is formed like a half-circle – a design that allows them to connect with each other like a jigsaw puzzle although, unlike a puzzle, nylon straps hold the blocks together. To test the TecPro concept, the containers were filled with polyethylene foam – a substance known to absorb high energies. Impact tests were conducted using a front nose similar to an F1 car, but the tests were unsuccessful due to the layers tearing again. To prevent the nose penetrating the material, two 2-mm steel sheets were then placed vertically in the center of each container, with 30-cm layers of foam either side. Once again, a nose-cone was crashed into the structure (Figure 5.9) at 127 km/hr and this time it stopped without any tearing or penetration.

Gramling's test was similar to the impact dynamics of a crash involving Felipe Massa during the 2002 Monaco Grand Prix. In that race, Massa hit the tire wall at Ste. Dévote corner at a similar speed and experienced similar deceleration without injury. Gramling figured he could combine the characteristics of the Ste. Dévote corner with the TecPro barrier to create a high-security F1 impact structure, and so the TecPro-Ste. Dévote hybrid was born. The Ste. Dévote barrier comprised TecPro blocks, a 2-m gap, four rows of tires, which had tubes of high-density polyethylene inserted into them, and finally a retaining wall. It was a robust configuration that withstood 173-km/hr impacts and deceleration forces of 65 G. They were impressive impact data which impressed the FIA and, in 2006, the barriers were first used at an F1 circuit at the Italian Grand Prix in Monza.

Crash testing

Another data source used to help F1 engineers to design more impact-tolerant cars is crash testing. F1 cars must pass several impact tests which are carried out under FIA guidelines at the FIA Institute's F1 Accident Reconstruction Facility (ARF). The ARF is designed to recreate the forces generated in all major impacts in F1 and to measure their effect on the driver. The rig and chassis (Figure 5.10) can handle impact forces of up to 100 G, which is the equivalent of a race car going from 100 km/hr to zero in one-tenth of a second. Put simply, the facility can recreate impact conditions that stretch to the fastest and most devastating crashes in F1.

The chassis depicted in Figure 5.10 can be positioned on the rig at any angle to replicate all conceivable front, side, or rear impacts. When there's an accident of particular interest, the FIA Institute can put a dummy in the chassis, reconstruct the "crash pulse" of the accident, and measure the dummy loads to see how the

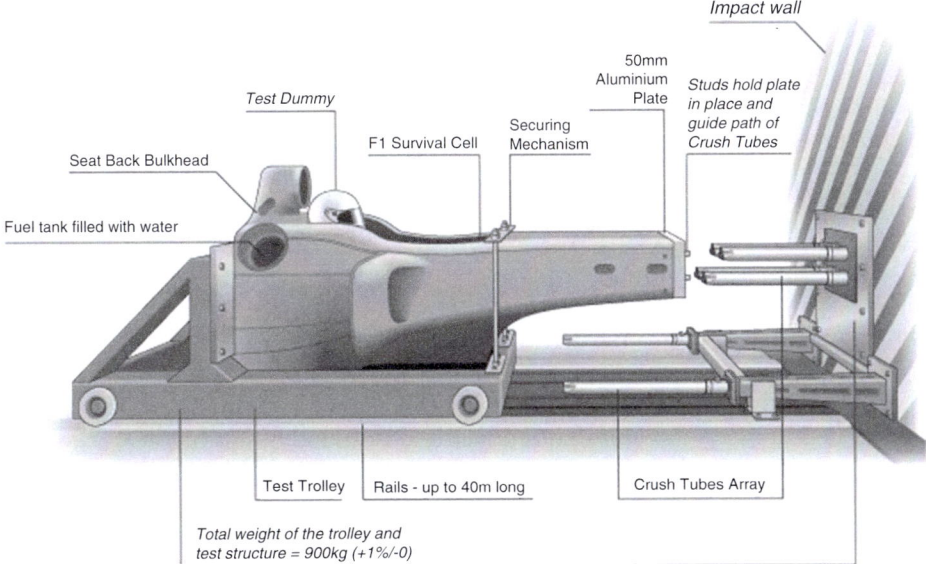

COMPONENTS OF THE FRONTAL IMPACT TEST

Figure 5.10 Infographic showing the components of the frontal impact test. Image courtesy: www.formula1.com.

whole system works. The reconstructed crash pulse data help engineers to determine how things like the headrest, belts, HANS device, and other systems contribute to the safety of the driver. The crash-test dummy is instrumented with accelerometers and transducers, which generate myriads of data such as chest acceleration and neck loads, lumbar forces, and bending loads in the thoracic spine.

The F1 crash-test dummy is very different from that used in a road-car test because the seating angle of an F1 driver is reclined, so the dummy has a special moveable pelvis to replicate this unusual posture. Once the dummy is loaded into the chassis, the rig is fired down the track at up to 100 km/hr before being abruptly stopped by a series of crush tubes that are designed to shape a pre-determined pulse. The configuration is basically a watered-down version of Colonel Stapp's research. Ten steel and aluminum tubes are used, each having a different shape and force. They can be configured depending on the crash pulse that is being replicated. Using the crash-test rig, F1 cars undergo front, side, and rear tests, after which engineers inspect the car to ensure that all structural damage has been limited to the car's impact-absorbing structures such as the side pods and the nose. Another element that undergoes particular scrutiny is the steering column, which must also pass an impact test. With so many instruments, it's not surprising that deceleration, energy absorption, and deformation loads are precisely defined. In the frontal test, for example, the deceleration measured on the chest of the dummy must not exceed 60 G within three milliseconds of the impact. In 2010, the frontal impact test became even more important following the ban on refueling during F1 races. That meant F1 teams had to increase the size of their fuel tanks so they were capable of holding 230 liters instead of the 100 liters that they previously held. The increase in the amount of fuel obviously meant that, in the event of a frontal impact, far greater forces would be experienced through the chassis than in previous seasons, which was why the FIA devised a new test.

It's often said that the job of a modern F1 driver is similar to a pilot of a fighter jet. After all, both occupations face similar G-loads – while fighter pilots withstand higher G levels, these loads are aligned only on a vertical axis, whereas an F1 driver endures these loads almost at right angles to the spine. Then there are the inherent similarities in driving around a circuit at speeds in excess of 320 km/hr and flying a jet at Mach speed. In terms of acceleration and deceleration forces, whether anticipated or unanticipated, the world of the F1 driver and the fighter pilot is a dangerous one. While F1 is much safer nowadays, it will be impossible to eliminate the danger completely. Accidents are part and parcel of F1, but the absolute aim has to be making them survivable; the development of deformable crash barriers, carbon-fiber survival cells, and indestructible helmets has gone a long way towards achieving that.

6 **Punching Out**

Ejection forces

Among professional pilots, there's a well-worn saying: "Flying is simply hours of boredom punctuated by moments of stark terror." It's an adage that Lockheed test pilot Bill Weaver can relate to. In January 1986, Weaver, together with flight-test specialist Jim Zwayer, was evaluating systems on an SR-71 (Figure 6.1) during a Blackbird test from Edwards Air Force Base.

Bill Weaver

Weaver and Zwayer were also investigating procedures designed to reduce trim drag and improve high-Mach cruise performance, which involved flying with the center of gravity (CG) located further aft than normal, thereby reducing the Blackbird's longitudinal stability. They had taken off from Edwards at 11.20 am and had completed the mission's first leg without incident. After refueling from a KC-135 tanker, the crew turned eastbound, accelerated to Mach 3.2 cruise speed, and climbed to 23,500 m, their initial cruise-climb altitude. A few minutes into cruise, the SR-71's right engine inlet automatic control system malfunctioned. The Blackbird's inlet configuration[1] was automatically adjusted during super-

---

[1] The main purpose of an inlet cone (sometimes called a shock cone) is to slow the flow of air from supersonic-flight speed to subsonic speed before it enters the engine. At supersonic-flight speeds, a conical shock wave forms at the apex of the cone (similar to the bow wave on a ship) and air passing through the conical shock wave slows to subsonic speed. The air then passes through a strong normal shock wave, within a diffuser passage, and exits at a subsonic velocity. This would normally be performed automatically but, in the early years of the SR-71 program, the analog air inlet computers couldn't always keep up with rapidly changing flight environmental inputs; if internal pressures became too great and the spike was incorrectly positioned, the shock wave would suddenly blow out the front of the inlet, with the result that the flow of air through the engine compressor would immediately stop, thrust would drop, and exhaust gas temperatures would begin to rise – due to the tremendous thrust of the remaining engine pushing the aircraft asymmetrically, this would cause the aircraft to yaw violently.

Figure 6.1 The Lockheed SR-71 was an advanced, long-range, Mach 3+ strategic reconnaissance aircraft developed as a black project in the 1960s by the Lockheed Skunk Works. Crews flying the SR-71 at 24,000 m faced two main survival problems: maintaining consciousness at high altitude and surviving ejection. Image courtesy: USAF.

sonic flight to decelerate airflow in the duct to ensure air reaching the engine's face was subsonic. This was achieved by the inlet's center-body spike translating aft and by modulating the inlet's forward bypass doors. Without the automatic adjustment, disturbances inside the inlet resulted in shock waves being expelled forward – a phenomenon known as an "inlet unstart". This caused an instantaneous loss of engine thrust, explosive banging noises, and violent yawing of the aircraft. Unstarts were not uncommon early in the SR-71's development, but a properly functioning system would recapture the shock wave and restore normal operation. On Weaver's planned test profile, the unstart had occurred on the right engine as the SR-71 entered a programmed 35° bank turn to the right. Instead of banking to the right, the aircraft *rolled* to the right and started to pitch up. Weaver jammed the control stick as far left and forward as it would go but there was no response. He explained the problem to Zwayer and told him they were in for a rough ride. In an attempt to solve the problem, Weaver explained that his plan was to stay with the aircraft until they reached a lower speed and altitude because he wasn't optimistic about their chances of surviving an ejection at Mach 3.2 at 23,500 m. Unfortunately, the G-forces had built up so quickly that Weaver's words were unintelligible to Zwayer. Worse, the cumulative effects of system malfunctions, reduced longitudinal stability, increased angle of attack, supersonic speed, and high altitude imposed forces on the airframe that exceeded the Stability Augmentation System's ability to restore control. For the crew, the events seemed to unfold in slow motion. As G-forces increased dramatically, the SR-71 started to disintegrate around them and Weaver blacked out.

Gradually regaining consciousness, Weaver was convinced that he was having a bad dream and would surely wake up. But, as he gradually became aware of rushing air and the sound of flapping straps, he quickly realized that he wasn't dreaming. He had separated from the airplane, although he had no recollection of initiating ejection. Making matters worse was the fact he couldn't see anything because his pressure suit's faceplate had frozen over.

Weaver was staring at a layer of ice.

He tried to figure out what had happened. He sensed he was falling and that his pressure suit was inflated, so he knew the emergency oxygen cylinder in the seat kit attached to his parachute harness was working. His next concern was trying to get stable because he was tumbling badly – as air density at high altitude is insufficient to resist a body's tumbling motions, centrifugal forces can increase rapidly and cause physical injury. It was for this reason that the SR-71's parachute system was designed to automatically deploy a small-diameter stabilizing chute shortly after ejection and seat (Figure 6.2) separation. But, since Weaver hadn't intentionally activated the ejection system, and because automatic functions depended on a proper ejection sequence, Weaver figured the stabilizing chute might not have deployed. Fortunately, it didn't take long before he was falling vertically. Weaver reckoned the stabilizing chute must have deployed, so his next concern was the deployment of the main chute, which was designed to open automatically at 4,500 m. As he plummeted to the ground, Weaver had no idea how high he was because he still couldn't see anything through his ice-encrusted faceplate and he had no idea how long he had blacked out. Concerned that the main chute might not deploy, he fumbled for the manual activation D-ring on his chute harness but, with his suit inflated and his hands numbed by cold, he couldn't find it. He was about to open his faceplate when he felt the reassuring deceleration of the main chute deploying. Breathing a sigh of relief, he raised the iced-over faceplate and saw that he was descending through clear, winter sky. About a quarter-mile away, he could see Zwayer's chute and, a couple of miles in the distance, he saw the smoking wreckage of what had once been the world's most advanced aircraft.

Weaver tried to turn to look in other directions to see whether he recognized any landmarks but his hands were still numb from the freezing temperatures. He was unsure of his whereabouts because, at the time of the accident, the Blackbird

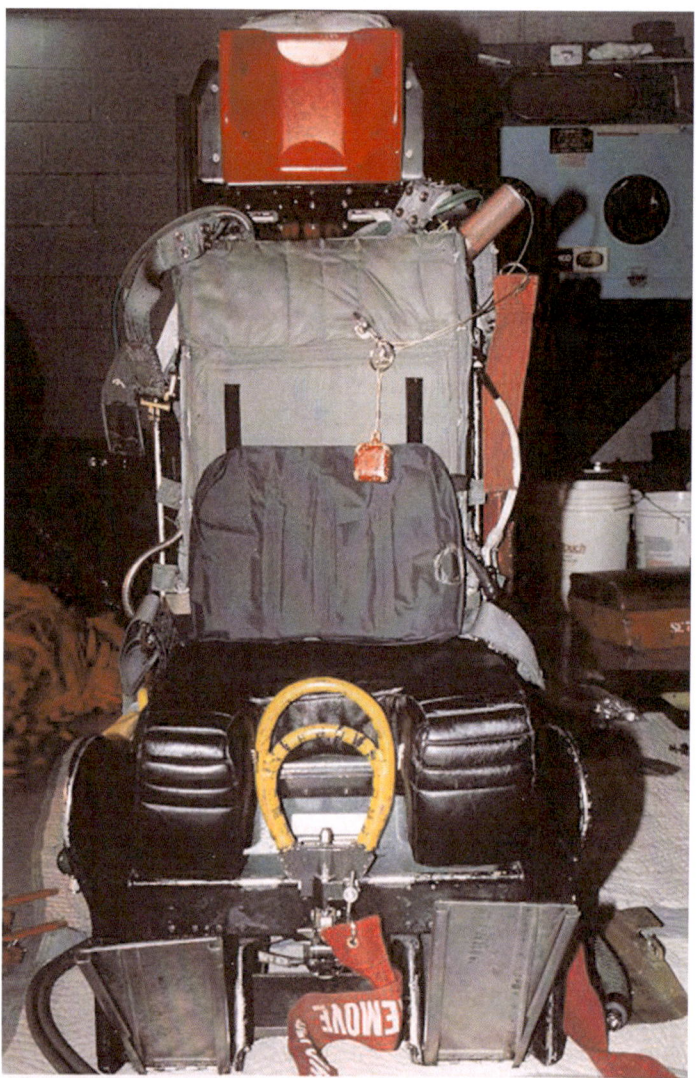

Figure 6.2 The SR-71 ejection seat was useable from zero speed and altitude to the maximum speed and altitude of the aircraft. The seat was a rocket-propelled, upward-ejecting unit. To eject from the SR-71, the crewman reached between his knees with both hands and gave a sharp, upward tug on the seat's large D-ring. After pulling on the D-ring, there was a 0.3-second delay to remove the canopy and then a catapult gas charge was fired to initiate seat ejection from the cockpit. The gas charge was sufficient to raise the seat above the canopy sills, at which point, a wire lanyard attached to the floor of the cockpit was pulled, igniting the seat's rocket motor. The rocket motor provided sufficient thrust and duration to eject the seat approximately 100 m above the aircraft. Image courtesy: USAF.

was performing a turn over the New Mexico–Colorado–Oklahoma–Texas border area. At Mach 3, the SR-71's turning radius was about 160 km, so Weaver wasn't even sure which state he was in. Preparing himself for landing, he deployed the seat kit's release handle to make sure he didn't injure himself when he hit the ground. Looking down, Weaver prepared for his first-ever parachute landing, hoping he could avoid the numerous rocks and cacti bushes that seemed to be everywhere. Fortunately, his landing was uneventful and he landed on soft ground. He quickly collapsed his chute and began to check his equipment. His lap belt and shoulder harness were still draped around him but had been shredded due to the extreme forces of Weaver having been ripped out of the aircraft (the ejection seat had never left the SR-71). He also noticed that the two oxygen lines that fed his pressure suit had come loose and were barely hanging on. If either line had disconnected at high altitude he would have died from rapid decompression. Having inventoried his equipment, Weaver looked at the ranchland around him.

A man was walking towards him wearing a cowboy hat. In the distance, a helicopter sat idling. Weaver thought he must be hallucinating. Even if he had been at Edwards and had informed the search-and-rescue unit that he was bailing out, the rescue team couldn't have gotten to him as fast as the guy in the cowboy hat had.

The guy in the cowboy hat was Albert Mitchell. He was the owner of a large cattle ranch in north-east New Mexico and when pieces of airframe had started raining down on his property, he became understandably concerned and went to investigate. Using the ranch's Hughes helicopter, it wasn't long before he spotted Zwayer. Then he saw Weaver's parachute. He landed near the chute and went to investigate. Weaver explained that he was part of a two-man crew and asked whether Zwayer had survived. Mitchell climbed into his helicopter and flew a short distance to Zwayer's body before returning to tell Weaver that his flight-test engineer was dead, his neck having been broken during the aircraft's disintegration. After Weaver checked on Zwayer's body, Mitchell flew him to Tucumcari Hospital, 100 km away. Mitchell had already notified the New Mexico Highway Patrol, the Air Force, and the hospital, and felt duty-bound to get Weaver to medical care as soon as possible, despite the Blackbird's pilot urging him to take his time. In considerable pain, Weaver couldn't help noticing the Hughes's airspeed indicator creeping beyond the "Velocity Not to Exceed" line as Mitchell tested the red-line capabilities of the ranch helicopter. As Mitchell gunned the Hughes to the hospital, Weaver, having just survived the fastest and highest ejection[2] in history, couldn't help pondering on the irony of dying in a

---

[2] In case you're wondering how Weaver could have survived an ejection at such high speed, bear in mind that, at the high altitudes that the SR-71 flew, the air pressure is significantly lower, thereby lowering the dynamic pressure. For example, an SR-71 flying at 3,200 km/hr at near 25,000 m is actually experiencing wind force (dynamic pressure) equivalent to about 460 mph (400 Knots Equivalent Airspeed (KEAS)).

helicopter crash en route to the hospital! Fortunately, they made it to Tucumcari safely and Weaver was able to contact Lockheed staff at the flight-test office at Edwards, who were amazed to hear Weaver's voice, since they had assumed no one could have survived.[3]

Weaver's exit from the SR-71 is perhaps one of the most unusual and extreme bailouts from any aircraft, especially since the ejection seat that was designed to do the job didn't leave the aircraft. Instead, Weaver's departure was caused by excessive G-forces acting on the Blackbird's airframe. Fortunately, in most situations that require a pilot to bail out, the ejection seat does its job.

Ejection events

Pre-ejection

The time interval from the initial need to leave the aircraft until ejection is initiated is known as the *pre-ejection* phase. During critical phases of flight, such as take-off and landing, this phase can be extremely short and allow precious little time to prepare for ejection. However, in other in-flight emergencies, there may be enough time to make changes to increase the probability of successful ejection. For example, speed can be reduced to lessen the effects of windblast and flailing, harness straps can be tightened, and body position can be adjusted to reduce injury from the ejection forces, which, as we shall see, can be considerable.

Primary acceleration

Ejection forces are primarily in the upward direction (Figure 6.3). In this phase, the objective is to attain the greatest possible velocity over a specified period of time. The force, which causes the seat to move upward, ranges between 12 and 20 G. The incidence of spinal injury appears to increase markedly if the peak acceleration exceeds 25 G and if the rate of onset is greater than 300 G/second. Many factors will determine the actual G-value that an ejection seat will produce. For example, the propulsion device will be affected by temperature, the total weight of the occupant-seat assembly, the aircraft velocity and relative airspeed at the time of ejection, and the ejection altitude. These accelerative forces will also be influenced by the complex mechanical interaction between the pilot's body and its relationship to the seat as well as how various body

---

[3] An investigation into the accident revealed the nose section of the aircraft had broken off aft of the rear cockpit and crashed about 16 km from the main wreckage. The debris field was about 25 km long and 16 km wide. Extremely high air loads and positive and negative G-forces had ripped Weaver and Zwayer from the aircraft. Two weeks later, Weaver was back in an SR-71.

Figure 6.3 Captain Christopher Stricklin ejects from his F-16 aircraft with an Advanced Concept Ejection Seat (ACES) II on September 14th, 2003; Stricklin was not injured. Image courtesy: USAF.

parts relate to each other. Given the dynamic way in which a human behaves during the ejection sequence, it can be analogized to a fluid-filled body and, like any fluid-filled body, there are maximum forces that can be imposed before limits are reached. For example, compression forces may be initially elastic and within the body's absorption ability but, in some instances, these forces may exceed the elastic limits and thus become "dynamic overshoots", which are important when considering the sorts of injuries sustained during the ejection sequence.

Windblast

After the initial acceleration of the seat going up the rails and the differential acceleration of entry into the airstream, the aircrew–seat combination is rapidly decelerated due to ram air force from windblast. This force is known as the Q-force and varies with the density of the air (this is the reason why the SR-71 seat was a low-Q seat) and is proportional to the surface area of the aircrew–seat combination. Q-forces are related to indicated airspeed rather than true airspeed and increase with the square of the velocity, which is why pilots are trained to *reduce* airspeed and *increase* altitude prior to ejection.

In ejection events, Q-forces have been divided into those produced by *windblast*, resulting in injuries such as petechial and subconjunctival hemorrhage, and those forces that cause injuries produced by *flailing* of the head and extremities. Flail injuries are caused by the differential deceleration of the extremities and their interaction with the torso and seat. A typical flail injury is caused by the extremities' leaving their initial position, building up substantial acceleration, and then suddenly stopping. A sudden stop may produce a bone fracture, joint dislocation, or total disarticulation, depending on the strength of the ejection forces.

Parachute descent and landing

This phase is critical to the outcome of the entire process of escape, yet 90% of all non-fatal injuries (Table 6.1) following a successful ejection occur during landing. Although the techniques of landing by parachute are easily taught and simulated by jumps from training towers, the incidence of sprained or fractured ankles is estimated to be 50 per 1,000 descents. The correct procedures for parachute landing are taught to aircrew during several phases of their training, but all the training in the world can't help prevent injuries if the parachute-opening shock is particularly severe, if the drogue chute fails, or if the main parachute deploys prematurely. High-altitude escape is relatively rare but, as occurred in the SR-71 bailout, if it *does* occur, there are additional risk factors the crew must be concerned with. For example, opening shock may be heightened due to increased velocities to the point that damage to the parachute and injury to the crewmember may result.

Not surprisingly, given the huge compressive forces that are transmitted along the spine (Figure 6.4a and 6.4b) during an ejection, back injuries have been the major source of reported harm during emergency egress from aircraft. While the vertebral column is able to absorb large compressive loads applied parallel to the long axis of the column, dynamic loading produced at angles of 12°–15° to the long axis of the spine is another matter and often results in significant injuries. The greatest risk occurs between the thoracic vertebral region, T10, and the lumbar region, L2 (Figure 6.4c). Of all injuries reported, compression fractures tend to be the leading type. Obviously, ejecting from an aircraft is dangerous enough without having to worry about the possibility of back injuries, which is

Table 6.1. Injuries occurring during and following ejection.

Event	Cause	Injury
Ejection	Ejection-seat G-forces	Spinal compression fracture
	Struck by seat or cockpit object	Extremity fracture
	Impact canopy structure	Severe lacerations
		Neck strains
		Spinal compression fractures
	Windblast	Petechial, retinal, and conjunctival hemorrhages
		Fractures
	Flail (linear)	Dislocation/disarticulation of extremities
		Internal injuries to organs
		Unconsciousness
	Accelerative and decelerative forces	Subdural hematoma
Parachute deployment	Parachute-opening shock	Cervical fracture of strain
		Muscle sprains
		Cervical vertebrae dislocation
		Lacerations
Parachute descent	High-speed rotation and/or spinning	Severe pain and hemorrhages
Landing	Landing impact	Leg-ankle fracture
		Spinal fracture

why ejection seats are carefully designed to minimize the stresses on the body as it is shot out of the aircraft.

Ejection seats

The Advanced Concept Ejection Seat (ACES) used by the United States Air Force (USAF) provides aircrew escape up to about 425 Knots Equivalent Airspeed (KEAS), although the performance limit is cited as 600 KEAS. However, very few successful ejections have occurred over 500 KEAS due to the effects of windblast.

Designed by the Douglas Aircraft Company, the third-generation ACES II has been in use since 1978. It ejects the aircrew in two stages. First, the entire canopy or hatch above the aircrew is opened or jettisoned, and the seat and occupant are launched through the opening. The ACES II performs both functions as a single action through the use of firing handles which activate the canopy jettison systems, followed by the seat ejection. Other aircraft designed for low-level flying sometimes have ejection seats that fire *through* the canopy because waiting for the canopy to be ejected is too slow. For example, the BAe Hawk uses a Canopy

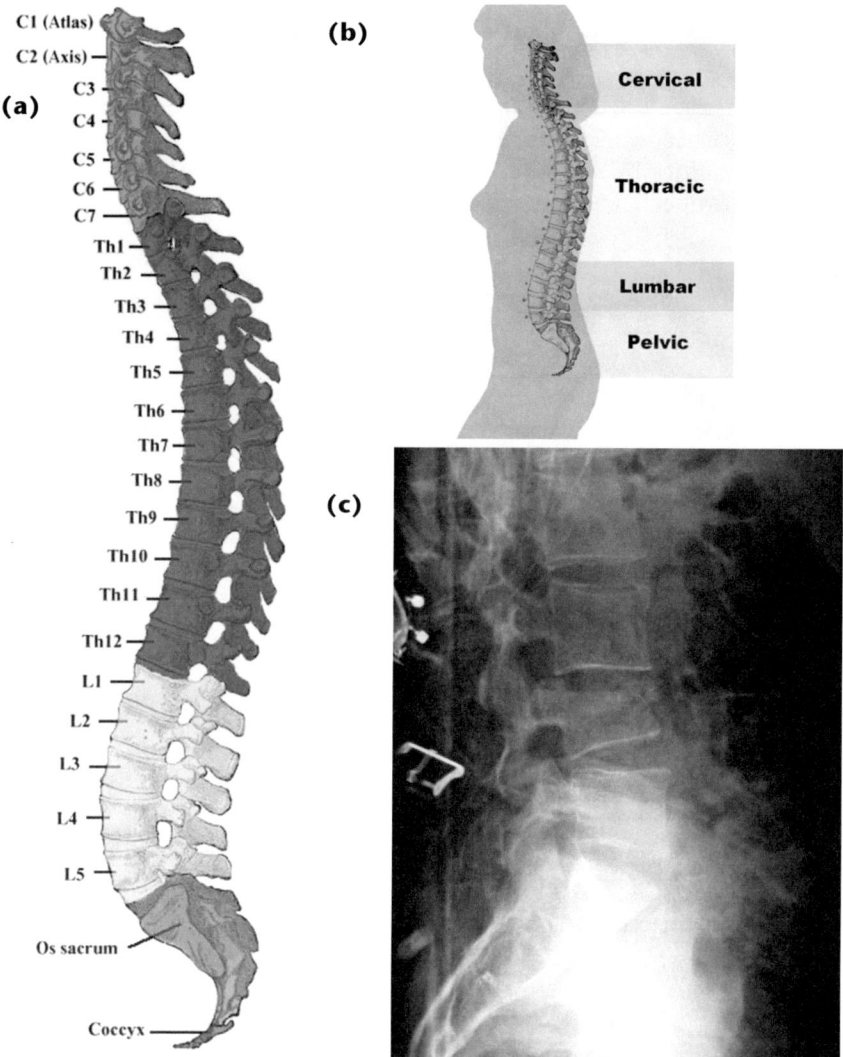

Figure 6.4 (a) The spine consists of 33 vertebrae arranged one above the other. The curvatures help in weight bearing by the spine. (b) The width/ diameter of the vertebrae varies, being maximum in the cervico–thoracic (neck and upper torso) junction and lumbo–sacral (lower back) region and minimum in the thoraco–lumbar (middle torso) region. The weight bearing of the spinal vertebrae increases from the neck downwards and at the lower-most thoracic (T-12) level, 60% of body weight is borne by this (T-12) vertebra. Any factor that deforms the normal curvatures of the spine reduces its tolerance to ejection forces. The most common cause of this is the forward bending of the spine thus reducing the vertebral surface area in opposition due to any reason and hence increasing the force per unit area. (c) An L4 compression fracture caused by ejection. Image courtesy: Wikimedia Commons.

Figure 6.5 BAe Hawk's Canopy Destruct system. Image courtesy: USAF.

Destruct system, which has an explosive cord (Miniature Detonation Cord (MDC) or Flexible Linear Shaped Charge (FLSC)) embedded within the acrylic plastic of the canopy (Figure 6.5). In this configuration, the MDC is initiated when the eject handle is pulled and it shatters the canopy over the seat a few milliseconds before the seat is launched.

Another common ejection seat is the Mk. 10, manufactured by Martin-Baker, the worlds longest-established and most experienced ejection-seat manufacturer. The Mk. 10 seat (Figure 6.6 and Table 6.2) is designed in four main units: catapult, main beam structure, seat pan, and parachute assembly – a manufacturing approach that simplifies and speeds maintenance and allows for rapid installation, assembly, and disassembly of the seat.

Operation of the Mk. 10 follows a dynamic sequence of events. First, the seat-firing handle is pulled, causing the seat initiation cartridge to fire. Next, the harness retraction unit is operated and the primary cartridge is fired, causing the inner and intermediate pistons to rise, releasing the top latch. The seat then rises up guide rails and the miniature detonating cord trip initiates the canopy-fracturing system. This event is followed almost immediately by a secondary cartridge firing as the seat rises and electrical connections separate. The aircraft portion of the main oxygen-generating system block separates, disconnecting the main and backup oxygen, after which the anti-G suit hose disconnects and leg-restraint lines draw back and secure the aircrew's legs. The leg-restraint lines become taut and rivets shear, freeing lines from floor brackets while barostatic

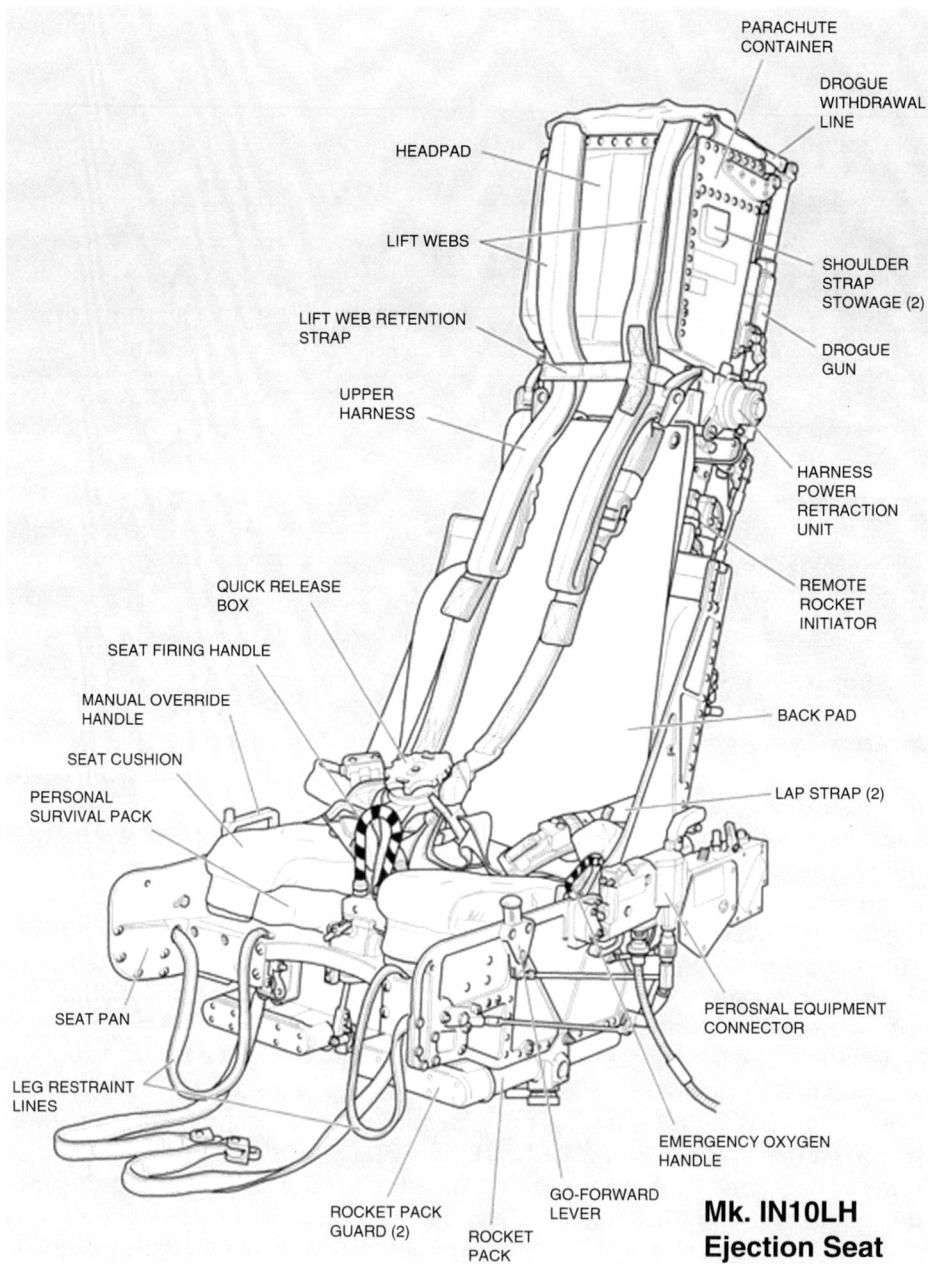

PARACHUTE
CONTAINER

DROGUE
WITHDRAWAL
LINE

HEADPAD

LIFT WEBS

SHOULDER
STRAP
STOWAGE (2)

LIFT WEB RETENTION
STRAP

DROGUE
GUN

UPPER
HARNESS

HARNESS
POWER
RETRACTION
UNIT

QUICK RELEASE
BOX

REMOTE
ROCKET
INITIATOR

SEAT FIRING HANDLE

MANUAL OVERRIDE
HANDLE

BACK PAD

SEAT CUSHION

PERSONAL
SURVIVAL PACK

LAP STRAP (2)

SEAT PAN

PEROSNAL EQUIPMENT
CONNECTOR

LEG RESTRAINT
LINES

EMERGENCY OXYGEN
HANDLE

ROCKET PACK
GUARD (2)

GO-FORWARD
LEVER

ROCKET
PACK

**Mk. IN10LH
Ejection Seat**

Figure 6.6 The Mk 10 seat is designed in four main units: catapult, main beam structure, seat pan, and parachute assembly. Image courtesy: Martin-Baker.

Table 6.2. Martin-Baker Mk. 10 ejection-seat specifications.

Operating ceiling	15,250 m
Minimum height/speed	Zero/zero in near-level attitude
Maximum speed for ejection	630 Knots Indicated Airspeed (KIAS)
Parachute type	GQ Type 1000 Mk. 2
Drogue parachute type	1.5 m and 0.55 m
Drogue deployment	Drogue gun
Harness type	Integrated
Ejection-seat operation type	Ejection gun and multi-tube rocket pack
Ejection gun	Single, two-stage
Ejection initiation	Handle on seat pan initiates gas-operated seat-firing system
Barostatic time-release unit	Yes, +G restrictor
Oxygen supply	Bottled emergency oxygen
Aircrew services	Personal equipment connector provides connections for: main oxygen, backup oxygen, emergency oxygen and anti-G suit
Command ejection	Yes
Canopy jettison	No
Miniature detonating cord	Yes

time-release units initiate. The emergency oxygen trips and the remote rocket initiator is operated by a static line as a cartridge fires to ignite the rocket pack which sustains upward thrust of the ejection gun, ensuring diverging trajectories for the front and rear seats. After the delay mechanism has operated, the drogue gun piston fires, the ejected piston withdraws the closure pin from the closure flaps of the drogue parachute pack, and the drogue chutes are deployed – an event that stabilizes the seat and aircrew. As the seat descends, the barostatic time-release unit operates to prevent parachute deployment above a pre-determined altitude; the barostatic-controlled G-switch delays parachute deployment above 2,100 m until speed and G-force are reduced. After the barostatic time-release unit operates, a normal parachute descent follows; the aircrew releases either of two quick-release connectors to lower the survival pack to the end of a line and waits for landing, after which the survival pack is opened.

It sounds fairly straightforward and effective, and it is. After all, thousands of aircrew have been saved thanks to the ACES II and Mk. 10. But the prospect of ejecting from an aircraft is becoming ever more challenging for aircrew as aircraft become more and more maneuverable thanks to advances such as progressive control systems and thrust vectoring features. Because of this, ejection-seat engineers have had to go back to the drawing board to figure out how to ensure survival at higher G-onset rates and velocities.

Future ejection systems

One of the problems faced by aircrew ejecting from high-performance aircraft such as the F-22 is simply being restrained by the harness system. To help aircrew during high-G ejections, the harness system has seen several changes from the conventional five-point harness system to the newer concept of the integrated harness assembly. The integrated harness assembly consists of lower and upper-torso restraint systems along with a Passive Arm Restraint System (PARS), which is activated on ejection. The lower-torso restraint system provides protection during normal flying maneuvers whereas the upper-torso restraint comes into play when the crew is exposed to a maximum of 30 G in the forward and downward axis. Thanks to the design of the harness assembly, the distribution of the ejection forces, crash loads, and parachute-opening shock are distributed evenly over a larger area of contact than conventional harnesses. Another key component that helps restrain the pilot is the Powered Inertia Restraint Device (PIRD). The gas initiation mechanism of the PIRD, commenced on ejection, helps restrain the pilot in the seat prior to ejection. It's a concept that has been introduced in the newer ejection seats.

While the integrated harness assembly and the PIRD do a good job restraining the pilot, these new features only help solve part of the high-G ejection problem, since there is still the issue of stabilizing the seat once it leaves the aircraft. The problem of stabilization becomes more pronounced the greater the G, which is why ejection-seat engineers have come up with the Directional Automatic Realignment of Trajectory (DART) system. The DART provides a stabilizing force during the rocket-burn stage based on the CG and aerodynamic movements. The DART system is composed of two lanyards, a disc brake, and a DART bridle. As the seat moves up, the lanyard plays out under minimal line tension and, once fully stretched, helps rotate the bridle downward for realignment of trajectory.

Assisting the DART in stabilizing the seat during the ejection is the stability package, or STAPAC. The STAPAC uses a gyroscopic mass and a set of pulleys to rotate a smaller rocket, which provides thrust in the opposite direction to the pitch rotation to halt the adverse pitch as quickly as possible. The STAPAC is primarily effective in the lower speed ranges, below 300 knots. At higher speeds, the drogue parachutes provide the required stability together with the DART.

Another advance in ejection-seat development is the increase in the level of automation. For example, the Martin-Baker Mk. 16 ejection seat, designed for the Joint Strike Fighter (JSF), is microprocessor-based controlled and features a suite of sensors that continuously update the various parameters of flight, such as attitude, altitude, airspeed, and even environmental conditions. It's a level of automation that is designed to improve the odds of survival and expand the boundary limits for successful ejection. The ability of a seat such as the Mk. 16 to monitor environmental factors has allowed better control inputs, thereby improving seat stability in high-G ejection scenarios.

Emergency egress for astronauts

When we think of escape pods for astronauts, we often imagine them jettisoning off Star Destroyers or the Battlestar *Galactica*, but they're not all fictional. After all, leaving astronauts without a means of escape in the event of an emergency would be very bad planning. In the heyday of the space race, if something did go wrong during a launch and the rocket was set to explode, there was another, smaller emergency rocket attached to the capsule (NASA and the Russian Space Agency utilized similar systems). The capsule would detonate a series of explosive bolts that detached the emergency rocket from the exploding rocket and would send the astronauts sideways away from the main rocket's explosion. Then, when the capsule was far enough away, more explosive bolts would explode, detaching the capsule from the main rocket's escape rocket. Despite this system reading like a Michael Bay movie, the one time a Launch Abort System (LAS) was actually needed, it was a complete success.

Well, nearly a complete success.

Soyuz T-10-1

On September 26th, 1983, cosmonauts Vladimir Titov and Gennadi Strekalov were preparing for their flight, Soyuz T-10-1. The Soyuz T-10-1 mission was to visit the Salyut 7 space station, which was occupied by the Soyuz T-9 crew, but it never finished its launch countdown. Just 90 seconds before launch, a fuel valve at the base of the rocket malfunctioned, opening and spilling fuel onto the launch pad. A fire broke out and flames engulfed the rocket with its 180 tonnes of highly flammable fuel. From years of training, Titov and Strekalov knew what would follow. The automatic launch-escape system would kick in and explosive bolts would fire, flinging the Soyuz T capsule free of the three-stage rocket. One second later, solid-fuel engines in a tower attached to the top of the capsule would ignite, lifting the Soyuz T orbital module and descent module away and clear. Five seconds after that, more explosive bolts would fire to separate the manned descent module from everything else. Its parachutes would release and its retro-rockets would fire, slowing the capsule enough for a safe landing.

Unfortunately, none of that happened.

The automatic launch-escape system didn't kick in because the fire had burned the system's wiring, preventing it from being activated automatically. Feeling strange vibrations and seeing black smoke and yellow flames outside their window, Titov and Strekalov tried to fire the launch-escape system manually, but there was no response. Their only hope was Mission Control. To fire the escape system manually from Mission Control required two operators, located in separate rooms, to press separate buttons at the same time. With flames rising from the launch pad and the rocket already leaning alarmingly 20° to the side, controllers scrambled frantically to get the system to free. Just 10 seconds after the flames first appeared, controllers managed to activate the escape system. The escape system motor fired, dragging the orbital module and descent module, encased within the upper shroud, free of the booster, throwing Titov and Strekalov more than 1,000 m into the air, during which they were exposed to almost 17 G of acceleration. Four paddle-shaped stabilizers on the outside of the shroud opened. The descent module separated from the orbital module at an altitude of 650 m and dropped free of the shroud. It discarded its heat shield, exposing the solid-fuelled landing rockets, and deployed a fast-opening emergency parachute. Then, the engines cut out, the descent module separated, and its parachutes unfolded. A moment later, the entire rocket and launch pad exploded. The blast was so intense that the capsule, now 5 km away, was thrown sideways and launch-pad workers in underground bunkers felt the pressure wave.

Strekalov and Titov landed safely, their capsule hitting the ground with a hard bump that shook both men up but did them no damage except for some bad bruises. Rescuers quickly pulled them from the capsule then gave them a glass of vodka to calm their nerves as everyone watched the nearby launch pad crumble in flames and clouds of smoke. It took 20 hours to put the fires out. Years later, in an interview with the American History Channel, Titov claimed that the crew's first action after the escape rocket fired was to deactivate the spacecraft's cockpit voice recorder because, as he put it, "We were swearing".

LASs

Rockets like Russia's Soyuz, China's (Soyuz-derived) Shenzhou, and NASA's Saturn V Moon launcher famously had escape towers on top of their capsules. The reason? If any, or all, of the rocket stages below the crew exploded on the launch pad, then rockets in the tower would lift the capsule clear of the mayhem and allow it to parachute to the ground a safe distance away, which is what happened in the Soyuz T-10-1 incident. Despite its drawbacks, which we'll get to shortly, the escape-tower concept is still being used on current spacecraft. For example, the next generation of space explorers will fly aboard NASA's Orion Multi-Purpose Crew Vehicle (MPCV), which features a LAS positioned on a tower atop the crew module. Composed of solid-fuel rocket motors, separation mechanisms, and an adapter structure, the LAS (Figure 6.7) will provide escape capability for the MPCV crew from pad operations through ascent.

Now, you may be wondering, why bother with a LAS? After all, can't the astronauts simply eject in an emergency? Well, designing a spacecraft with ejection seats is not that easy from an engineer's perspective, which is why Titov and Strekalov didn't have an ejection capability. Simply stated, for every astronaut to safely bail out or be thrown clear to parachute height on a rocket-assisted ejection seat, they would each have to be sitting near an escape hatch, which means fitting lots of hatches – and fitting lots of hatches weakens a spacecraft's structure. That's not to say ejection seats haven't been used in spacecraft. The now-retired Space Shuttle used ejection seats during its first four test flights, but only for the two pilots up front – the seats were removed after the

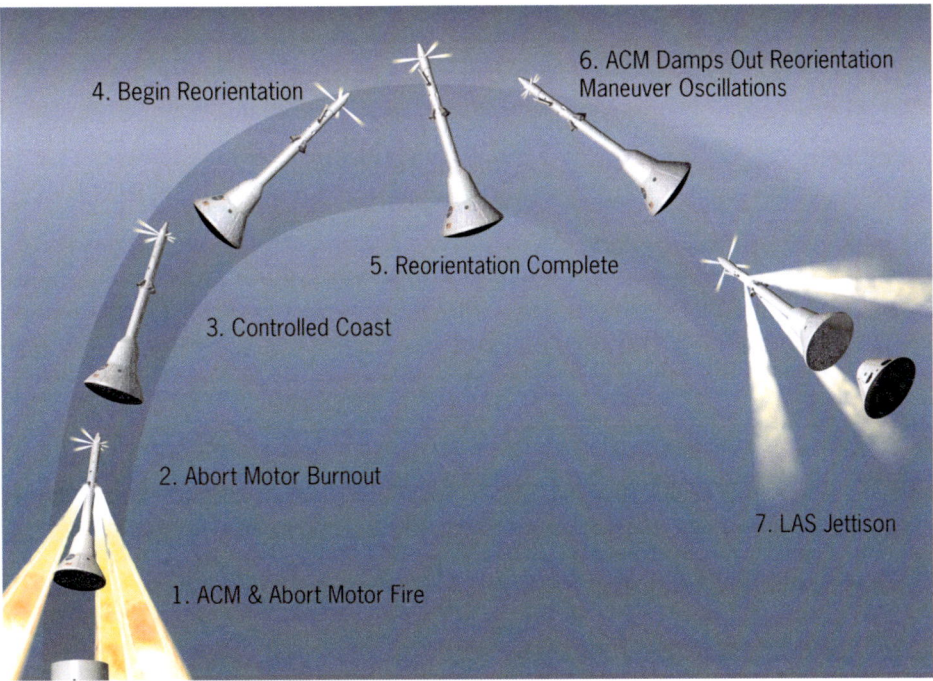

4. Begin Reorientation

6. ACM Damps Out Reorientation Maneuver Oscillations

5. Reorientation Complete

3. Controlled Coast

2. Abort Motor Burnout

7. LAS Jettison

1. ACM & Abort Motor Fire

Figure 6.7 Launch abort sequence. Image courtesy: NASA.

shuttle became operational (the ejection seats had limited use anyway, since they were only of use at speeds of up to Mach 3 and the top speed of the shuttle was Mach 18).

The 11-G LAS

The modern-day NASA LAS comprises three solid-propellant rocket motors. The first, the primary motor, is the abort motor, which, in the event of a pad abort, would fire nearly 500,000 pounds of thrust to propel the capsule away from the pad. The second motor, the attitude motor, would exert up to 7,000 pounds of steering force to maintain stability and reorient the vehicle as required, while the third motor, the jettison motor, would pull the whole LAS away from the crew module and make way for parachute deployment and landing. In the event of an abort (Table 6.3), the abort motor would burn for about six seconds, during which the crew module would reach 725 km/hr in the first three seconds (subjecting the crew to 11 G) in its upward trajectory away from the pad and an altitude of more than 1,500 m. As the abort motor burned out, explosive bolts would fire and the jettison motor would separate the spent abort system from the crew module to allow the parachute system to be deployed. Drogue parachutes would deploy first to stabilize the crew module through its initial descent and, after a few seconds, they would be cut away and

Table 6.3. Launch Abort System timeline.

T–0 Abort execute command	**L+25 seconds** Crew module drogue parachutes deploy
L+6 seconds Abort burn complete	**L+31 seconds** Drogue parachutes are cut away and three pilot parachutes are deployed
L+10 seconds Attitude control motor orients vehicle for LAS jettison and parachute deployment	**L+50 seconds** Pilot parachutes fully open
L+21 seconds Jettison motor separates LAS from crew module	**L+53 seconds** Crew module reaches descent on parachute
L+22 seconds Forward bay cover separates from crew module and two small parachutes deploy	**L+1 minute, 37 seconds** Crew module touchdown

three mortars would fire the pilot parachutes out into the air stream. After about 100 seconds under canopy, the crew module would land about 1.5 km downrange from the launch pad.

To try out the system, NASA tested the LAS at the White Sands Missile Range near Las Cruces, New Mexico. Called Pad Abort-1, the 97-second flight test, which took place on May 6th, 2010, was the first fully integrated test of the LAS developed for the MPCV. In the test, the crew module contained a dummy with some instrumentation. As advertised, the system accelerated to about 725 km/hr in just 2.5 seconds, subjecting the dummy to an acceleration of 16 G (a little higher than advertised, so it was a good thing it was just a dummy!). The parachute system worked like a charm, returning the module to the ground at 25 km/hr, which was about 10 km/hr slower than predicted.

New abort systems

While NASA's pad abort test was successful, the type of abort system tested is old technology. The problem with NASA's escape configuration is that the tower's weight slows the launch and, if it's not needed, it dumps a perfectly good set of rocket motors and fuel into the ocean. In short, it's a waste of resources. And, besides, there are better ways to do it. Leading the charge in the development of a new launch-escape system is wunderkind spaceflight outfit, SpaceX, which is designing a system for its forthcoming human-rated rockets. SpaceX's escape system doesn't require a clunky booster tower to be mounted on top of the crew capsule. Instead, SpaceX's seven-seater crew module (the Dragon) will integrate a number of escape rocket motors into the side walls of the spacecraft capsule, blasting downwards. This design not only drives the costs down (the Dragon is being developed as a reusable spacecraft), but also offers the crew an abort capability during the whole flight.

Another company involved in the design of a new escape system is aerospace giant Boeing, who have decided to use a system similar to SpaceX's for their upcoming seven-seater Crew Space Transportation capsule, the CST-100. Boeing's CST-100 is a general-purpose capsule that could be launched on the Atlas V, Delta IV, or even SpaceX's Falcon 9 rocket by 2015. The "pusher" design, which will feature as the CST-100's escape system, will use escape thrusters mounted beneath the capsule to eject the crew clear in the event they are faced with a Soyuz T-10-1 scenario. The drawback? If unused, the emergency fuel adds to the payload weight taken to orbit. But at least the motors are not thrown away.

7   Launch and Re-Entry

Ensconced inside the crew capsule, the astronauts check their straps one more time, listen to the endless series of acronyms being checked off by the flight controllers, and watch the commander follow the checklist. In just a few minutes, the vehicle will lift off from the pad in a thunderous cloud of fire and smoke.

Two minutes before launch, Mission Control reminds the crew to close their visors – a command confirming launch is imminent. The crew oblige, snapping the bail into place to seal the faceplate. On their left knees, the crew flips a lever on their suits to start the flow of oxygen. With adrenaline pumping and their heart rate close to the red line in anticipation, the crew tries to slow their breathing and control the attack of butterflies. For a moment, the thought that these could be the final moments of their lives enters their head, but their attention is quickly drawn to the expectancy of the rocket lighting up. Bodies tingling with anticipation, the astronauts give each other a thumbs-up.

With one minute to launch, the primary flight displays continue scrolling data as the astronauts think about their families and all the training and preparation that brought them to this pivotal moment in their lives. Mission Control announces 10 seconds to launch – a command indicating the ground launch sequence has taken over the count. Next is the command: "GLS is Go for auto sequence start." At 15 seconds to go, the attitude indicators on the primary flight displays scroll to the correct launch attitude as the vehicle processes its final update.

At T minus 5 seconds, tonnes of water flood the pad and cascade into the flame trench designed to absorb the shock waves of ignition. Finally, the countdown reaches the endpoint. Ignition commands from the vehicle's flight computers commit the vehicle to flight as electrical energy is routed from the first-stage avionics to the motor igniter. The igniter initiates the burn of the five-segment motor. As the motor burns, it builds up combustion along the surface of the propellant, which is expelled out of the nozzle, creating the thrust for lift-off. The pyrotechnic charges holding the vehicle to the pad explode and the vehicle shoots towards the sky. A rumble shakes the stack as the main engine spews fire and roars to full power. On the primary flight displays, the computers rapidly scroll through engine checks and mark the mission time: zero. The clock is running! Inside the capsule, the astronauts are pushed back into their seats by a force more than one and a half times the force of gravity. Mission Control makes the call: "Tower Clear!" The rocket is already consuming propellant voraciously at 5 tonnes/second, burning at a searing $2,760^\circ$C. The crew are pushed further back into their seats as the gravitational forces ramp up to nearly 3 G.

Less than 30 seconds after launch, shock waves form on the nose and the vehicle begins to vibrate. Meanwhile, the vehicle's control system continuously monitors trajectory and issues guidance commands to ensure the rocket nozzle's angle is correct. A few seconds later, the commander announces Mach 1 and the crew listen to the rushing sound of supersonic air being displaced by the vehicle. Quickly, the sky turns from blue to black, the cockpit becomes more silent, and the ride becomes noticeably smoother. The commander signals the crew to shut off the oxygen and open their visors. The vehicle is now more than two minutes into flight and flying at an altitude of more than 60 km. The astronauts feel another thump as the upper stage's engine ignites, propelling the vehicle to Earth orbit. A few more seconds pass and the commander announces an altitude of 100 km. Meanwhile, the rocket, reduced to its first stage and the capsule, continues to accelerate, the altimeter tape scrolling madly on the primary flight display. Mach 15. The rocket is tearing along at more than 5 km/second! Mach 18 – 6 km/second! With a force of more than 3 G pressing down on them, the combined weight of each crewmember and their flight suit is more than 300 kg. Twenty times the speed of sound! And still it continues to accelerate. Finally, the Mach meter reaches Mach 25, a velocity of more than 8 km/second, and the commander welcomes the crew to space.

In the course of the last 50 years, humans have risen above Earth's atmosphere, walked on the Moon, performed complex mechanical tasks in the vacuum of space, and spent uninterrupted months on board the International Space Station (ISS). The public has grown accustomed to seeing astronauts in space, to the point at which they take for granted the health and safety of the planet's spacefarers. There is little doubt that the success of international manned space programs is responsible for misleading the public into believing that space travel has few associated risks, but nothing could be further from the truth. From launch to landing, astronauts are exposed to myriad hazardous situations that have the potential to cause irreparable harm, many of which are rooted in the accelerative and decelerative forces encountered during lift-off and re-entry.

Astro-chimps

Imagine yourself strapped tightly into a seat ensconced in a claustrophobic metal cabin perched atop a rocket (Figure 7.1) full of explosive chemicals, waiting for those explosives to be ignited, after which you will be accelerated to such high speeds that you will be unable to move for several minutes. Welcome to the world of Chop Chop Chang, the chimpanzee.

Now you may be wondering what a chimp has to do with the space program but, to answer that question, we have to go back to the 1950s. As part of the space race against the Soviet Union, the Project Mercury program (1958–63) was tasked with putting an American astronaut into orbit and returning him safely. The program would also test how well humans functioned in the unknown

Figure 7.1 The Mercury-Redstone (MR) was an unmanned booster development flight in the US Mercury program. It was launched on March 24th, 1961. After problems encountered during the MR-2 mission carrying the chimpanzee Ham, it was apparent that the Redstone needed further development before it could be trusted to carry a human passenger. Image courtesy: NASA.

environment of space. But, before humans could be launched, NASA needed to make sure their astronauts could be kept safe from micrometeoroids, radiation, noise, vibration, microgravity, the vacuum of space, and, of course, those G-forces. Also, medical experts were unsure whether humans could handle being isolated and confined in a space as small as the claustrophobic interior of a space capsule. So, before taking a risk with a human, scientists recruited astro-chimps. Chimpanzees were an obvious choice because a chimp's organ and skeletal structures are similar to ours, and chimps can be trained. The first astro-chimp was launched on board a Mercury-Redstone 2 rocket on January 31st, 1961. Constructed of titanium just 0.01 inches thick, blanketed by fiberglass insulation and covered with blackened heat-radiating shingles, the rocket's sole passenger was a four-year-old ape born in Cameroon: Chop Chop Chang. Chop Chop Chang, or Number 65 as he was also referred to, was, by all accounts, a smart, loveable chimp with a positive temperament – ideal astronaut material in other words. Weighing 16 kg, Number 65's mission was to test the environmental control systems inside the Mercury capsule, to determine whether the Mercury-Redstone rocket was safe for humans and whether primates could function under the stress and pressure that come with space travel. Strapped inside his coffin-sized "cockpit", Number 65 was anything but a passenger, since he had a number of tasks to complete. For the correct response, he earned a banana pellet and, for a wrong response, he got an electrical shock. There is no indication whether NASA tried this type of training with their astronauts!

Number 65's flight demonstrated that chimps could concentrate and work in flight. Through the launch, more than six minutes of weightlessness, and re-entry, Number 65 moved levers in response to flashing lights, just as he had been taught in the laboratory. In fact, his response times in space were as good as on Earth. Unfortunately, the fuel in his rocket burned off too quickly and he was propelled almost 200 km further than planned. He also experienced acceleration forces up to 14.7 G, which was about 3.3 G more than planned. Making matters worse, Number 65's capsule made a rough landing downstream, the impact upon hitting the ocean's surface making his capsule begin to take on water. The landing bag and heat shield were designed to extend down about 1.2 m before landing and filling with air to help cushion the impact. Once landed, the bag and heat shield were supposed to act together like a sea anchor to keep the capsule upright. But a mistake at launch meant that the capsule went 65 km higher than planned and came down with such force that the heat shield punctured the capsule and water started to come on board.

Waiting to pick up the capsule were six naval destroyers and a landing ship dock, the USS *Donner*, with three helicopters on board. Unfortunately, because of the launch glitch, the task force was waiting in the wrong place. Just in case of such an event, four surveillance aircraft were ready to search for the capsule and, half an hour after the capsule's landing, a plane spotted it and helicopters from the USS *Donner* were sent to collect it. Once back on the ship, Number 65 was finally taken out of the capsule. He shook hands with the captain, ate an apple and half an orange, and was checked by a doctor who pronounced the chimp to be in good condition, despite the fact that his 16-minute flight into space had extended to a rescue mission lasting just under four hours. In fact, Number 65 showed no ill effects from his flight. Well, almost no ill effects. It seems Number 65, whose flight had earned him a new name – Ham, derived from his home unit, the Holloman Aeromedical Laboratory – wasn't too taken with the whole spaceflight experience because when the press wanted photos of him in his couch, he fought to avoid being strapped in (Figure 7.2). Nevertheless, the flight was pronounced a success and Ham[1] became a cause for celebration. He landed on the cover of *Life* magazine and was covered by all the newsreels. The Mercury astronauts were especially pleased that Ham had suffered no ill effects because they knew it wouldn't be long before one of them would be strapped on top of the Mercury-Redstone rocket (Alan Shepard would make America's first suborbital flight on May 5th, 1961, aboard the Freedom 7).

Now, you might have thought that Alan Shepard's blasting into sub-orbit would have signaled the end of primate flights, but Shepard's flight was a *suborbital* mission. What NASA needed now was proof that a primate could survive an *orbital* mission. Enter Enos. Unlike Ham, Enos was considered to be temperamental and a bit on the mean side, although the electrical shocks could have had something to do with that! Despite being a little highly strung, Enos was considered very intelligent and, for this reason, he was selected to make an orbital flight aboard a Mercury-Atlas rocket. The flight, which took place on November 29th, 1961, marked the final primate mission of Project Mercury. Like Ham's flight, Enos's mission wasn't without its problems. Enos's Mercury capsule malfunctioned and, for every correct move he made, he was given a shock instead of a banana pellet but, despite the discomfort he experienced, the chimp, ever the consummate professional, continued to make the correct moves. He splashed down in the Atlantic after more than three hours of flight, 181 minutes of which were in weightless conditions. Like Ham, Enos received a hero's welcome home but, while he made headlines, it was nothing compared to the star treatment accorded to John Glenn following his orbital flight in the Friendship 7 on February 20th, 1962. Glenn became an instant celebrity, as did all the Mercury astronauts. In his speech to Congress, Glenn said he was humbled when he met Caroline Kennedy and her first question was "Where's the monkey?".

Apollo G-forces

Fortunately for the Mercury astronauts, none of them was subjected to the G-forces that Ham or Enos had had to contend with. Mercury astronauts were subjected to maximum acceleration forces of between 6 and 7 G, whereas the Titan rockets that launched the Gemini astronauts subjected the crews to about

---

[1]   A 2008 animated film entitled *Space Chimps* was about sending chimps to space. The main character and hero of the movie was named Ham III, the grandson of Ham.

Figure 7.2 Ham fitted into a special biopack couch prior to flight into space. Ham (July 1956–January 19th, 1983), also known as the astro-chimp, was the first chimpanzee to be launched into space in the American space program. Image courtesy: NASA.

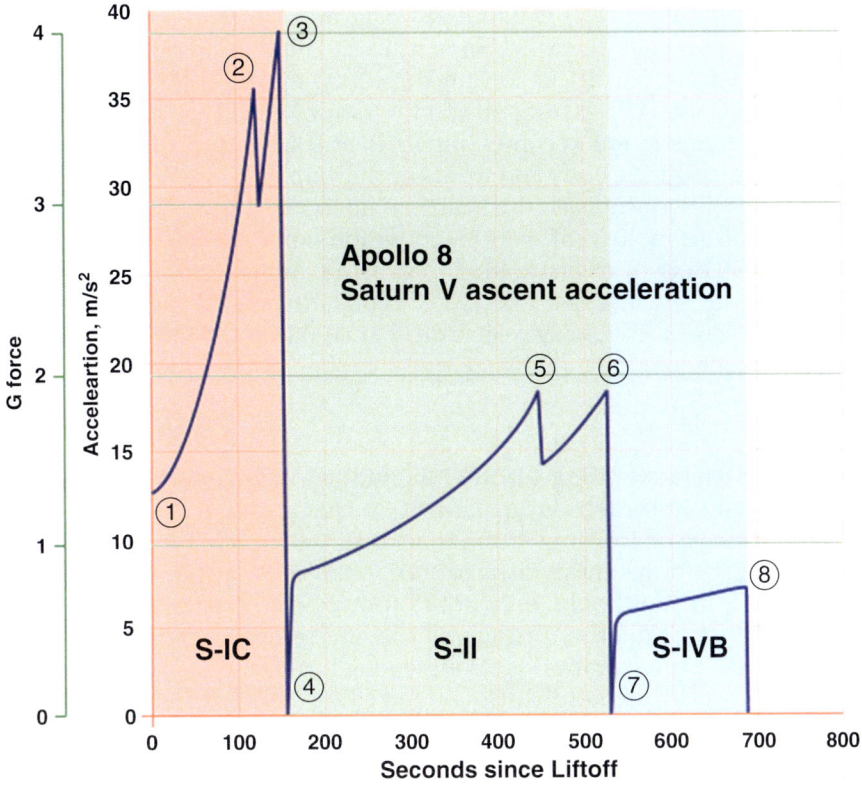

Figure 7.3 Apollo G-forces. Image courtesy: NASA.

7 G. The program that followed on from Gemini was the Apollo program, which used the Saturn V behemoth. The diagram in Figure 7.3 is derived from the AS-503 Saturn V Flight Evaluation Report and shows the G-forces experienced by the crew throughout the ascent.

At launch, the Apollo 8's first stage delivered more thrust than expected due to the launch vehicle being lighter than most of the later Apollo-Saturns. Therefore, compared to other missions, the graph in Figure 7.3 starts off at about 1.25 G (point 1 in the figure). For comparison, Apollo 15 left the launch pad at a statelier 1.1 G due to it being a relatively heavy vehicle with a first stage that showed a slightly lower-than-expected thrust. Going back to the graph, you can see at S-IC inboard engine cut-off (point 2) how steeply the acceleration is rising and, at S-IC outboard cut-off, the overall thrust and acceleration rise have been reduced, reaching a peak of 4 G by the time of S-IC cut-off (point 3). The next stage in the launch sequence was S-II ignition (point 4) followed by S-II cut-off of the center engine (point 5); outboard engine cut-off of the S-II occured at a peak of 1.8 G (point 6). The S-IVB ignition phase (point 7), which used a single J-2 engine, resulted in an acceleration of about 0.5 G, which gently rose for the duration of the burn to 0.7 G (point 8). Following S-IVB cut-off and orbital insertion, the spacecraft and S-IVB were in Earth orbit and weightless.

As we know from our discussion on the effects of Gs on pilots, the concern about the effect of Gs was rooted in physiology. We know that as G-forces increase, blood circulation becomes impaired, as the heart must work harder to pump blood through the body and we know this impaired circulation can lead to a dimming of consciousness, as the heart can no longer pump blood to the brain. The end result is a loss of peripheral vision and gravity-induced loss of consciousness (G-LOC), or blackout. For an astronaut engaged in the operation of a beast like the Saturn V, the effects of Gs encountered during launch could be deadly, but it wasn't just the acceleration that could cause G problems.

Pogo

Pogo is rocket scientist's slang for the longitudinal vibration or oscillation that sometimes occurs in rockets during launch. As the propellant flows through the pipes and fittings on its way from the tanks to the engine, low-frequency disturbances can form. These disturbances sound a little like the groaning of plumbing pipes in your house, only not as strong. Since this causes variations in the propellant flow rate, the thrust from the affected engines fluctuates several times a second. If the vibration is strong enough, it can have an impact on the crew. Vibration tests on astronauts, conducted by NASA at the Ames Research Center (ARC) at the dawn of the manned space era, confirmed human discomfort limits were reached at just 0.5 G. Pain was directly associated with motion of the eyeballs and testicles, as well as from internal heating that resulted from sloshing of the brain and viscera. Making matters worse, the vibration frequency was also in the range of normal brain waves, adding confusion to decision-making, hand and arm movement, and even speech. Worse was to come when flight data from the initial flights of the Titan II vehicle, which took place in early 1962, showed vibration levels as high as 5 G – a level much higher than the 0.5 G that was the maximum allowed. Hence, the program to launch a two-man Gemini capsule was jeopardized.

Not surprisingly, the unusual pogo phenomenon was quickly recognized as a dangerous threat to the whole endeavor of manned rocket flight and attempts were made to understand and resolve the problem. In an attempt to solve the pogo issue, engineers conducted extensive testing to generate test-verified equations describing the propellant feed systems and the engines. Fuel pump tests showed that, as inlet pressures were reduced towards cavitation,[2] the pump

---

[2]   Cavitation means that cavities/bubbles form in the liquid being pumped. These cavities form at the low pressure or suction side of the pump, causing several things to happen all at once: (1) the bubbles collapse when they pass into the higher regions of pressure, causing noise, vibration, and damage to components, (2) the pump can no longer build the same pressure, and (3) the pump's efficiency drops.

started acting as an amplifier, causing large oscillations in the thrust chamber pressure. By late 1963, the studies were complete. The fuel pump had been modified slightly, and a piston accumulator had been built for the fuel system and added to the vehicle. The system was flown with suppression devices on both the fuel and oxidizer feed lines, with unqualified success. With the pogo problem mitigated, the Titan II rocket was pronounced suitable for human flight but, just to be sure, NASA launched two more unmanned Gemini flights. Encouragingly, the post-flight data report from Gemini flights I and II showed insignificant pogo problems and the stage was set for manned flights.

The first Gemini crew of Virgil Grissom and John Young launched on March 23rd, 1965. The vehicle lifted off so smoothly that the astronauts remarked there was less noise and vibration than in the moving-base simulator in Dallas. Unfortunately, later crews weren't as lucky; the Gemini V crew of Gordon Cooper and Pete Conrad (Figure 7.4) reported pogo two minutes into

Figure 7.4 Gemini V was the third manned Gemini flight. Commanded by Gordon Cooper and piloted by Pete Conrad, the flight marked the first time an American manned space mission held the world record for duration, set on August 26th, 1965, by breaking the Soviet Union's previous record set by Vostok 5 in 1963. Image courtesy: NASA.

Figure 7.5 Apollo 6, launched on April 4th, 1968, was the Apollo program's last unmanned test flight of its Saturn V launch vehicle. It was intended to demonstrate full lunar injection capability of the Saturn V, and the capability of the Command Module's heat shield to withstand a lunar re-entry. Fuel line failures in several Saturn V second and third-stage engines prevented it from achieving lunar injection, but it was able to get close to lunar return velocity by using the Apollo spacecraft's engine, as was done on Apollo 4, the first Saturn V test. Despite the engine failures, the flight nonetheless provided NASA with enough confidence in the Saturn V to use it for manned launches. And, since Apollo 4 had demonstrated the heat shield at full lunar re-entry velocity, a potential third unmanned flight was canceled. Image courtesy: NASA.

their flight that made it impossible for them to not only read the panel gauges, but also to speak to each other. Fortunately, the Gemini V flight was the only one in the program to experience significant pogo.

In parallel with the Gemini Program, NASA was developing the Saturn rockets to take astronauts to the Moon and, with the outbreak of pogo on nearly all large launch vehicles, the Saturn V Moon mission had engineers concerned. The engineers were encouraged when no pogo was observed during the Saturn I flights and, when the first Saturn V launch, flight AS-501, occurred on November 9th, 1967, with no hint of a pogo problem, the engineers thought they might have solved the problem. But, during the AS-502 launch (Figure 7.5) on April 4th, 1968, big oscillations were seen on the accelerometer in the first-stage burn. The oscillations occurred at 105–140 seconds after lift-off, at an amplitude of 0.3 G at the aft of the vehicle, where the five F-1 engines were located. The pogo vibration wasn't severe enough to impair the structural integrity of the launch vehicle or spacecraft, but the amplitude was greater than that observed during the Gemini missions.

Not wanting to take any risks, NASA assembled a special pogo task force from NASA industry and academia for intensive investigation of the problem (at one time during the pogo studies, about 1,000 engineers were working on the problem). On engine test stands, inlet system pulsing tests were conducted on the F-1 engine to assess combustion-chamber vibration, and measures were implemented to resolve the problem before manned flights commenced.

On December 21st, 1968, the first manned Saturn V, AS-503, lifted off to demonstrate the capability of the Apollo 8 command module to perform the lunar mission. Following engine start, all five engines performed beautifully and the engineers had reason to be cautiously optimistic but, near the end of the second-stage burn, Frank Borman, the command pilot, reported he felt pogo. Data indicated that, 50 seconds prior to second-stage cut-off, the center engine had begun to vibrate at 18 Hz. Analysis showed that the crossbeam-mounted center engine (Figure 7.6) and the liquid oxygen tank bottom were vibrating like two prongs of a giant tuning fork.

The engineers decided that, by increasing the tank pressure, cavitation would be eliminated and pogo reduced. An alternate method was to turn the center engine off 75 seconds early and burn the remaining engines longer, with some reduced stage performance. The decision was to go with increased pressure, and flight AS-504 proceeded in about the same way as AS-503, although instrumentation showed G levels too close to structural limits – an event that resulted in subsequent flights having the center engine cut-off early. The next flight, AS-505, was nominal, but the third stage, Douglas S-IVB, showed a clear pogo oscillation of 20 Hz from 50 to 100 seconds into the burn of that stage. However, flight history showed that the oscillation was not a generic pogo problem and the occurrence was never repeated. On flight AS-506, the astronauts complained of high vibration during the second burn of the S-IVB that turned out to be the chattering of a non-propulsive vent valve system attached to the skin of the vehicle. Douglas changed the tolerance on the valves and that problem was

Figure 7.6 A close-up of Saturn V's F-1 engine. Developed by Rocketdyne under the direction of the Marshall Space Flight Center, the F-1 engine was utilized in a cluster of five engines to propel the Saturn V's first stage, the S-IC. Image courtesy: NASA.

resolved. Flight AS-507 showed four well-defined pogo occurrences during the S-II burn which looked like they might be nonlinearly limited. With flight AS-508 ready to go, decision time had arrived.

Rockwell's Space Division Rocketdyne, which had designed and built a pogo suppresser, felt it would eliminate the problem, but to install it would require technicians' crawling over the rats nest of cable harnesses to get into the center engine. Then there was the problem of inspection and verification, which would be just as difficult. So, the unanimous vote was to fly AS-508 as it was, just like AS-507. The engineers should have known better. After all, this was Apollo 13 (Figure 7.7).

Figure 7.7 The original Apollo 13 prime crew. From left to right are Commander James A. Lovell, Command Module pilot Thomas K. Mattingly, and Lunar Module pilot Fred W. Haise. Apollo 13 was the seventh manned mission in the American Apollo space program and the third intended to land on the Moon. The craft was launched on April 11th, 1970. The lunar landing was aborted after an oxygen tank exploded two days later, crippling the service module upon which the Command Module depended. Despite great hardship caused by limited power, loss of cabin heat, shortage of potable water, and the critical need to jury-rig the carbon dioxide removal system, the crew returned safely to Earth on April 17th. Image courtesy: NASA.

Apollo 13 launched on April 11th, 1970. During the second-stage burn, two episodes of pogo occurred on the center engine as expected from previous missions. So far, so good. Then a third pogo incident occurred that really caught the attention of the engineers. An acceleration at the engine attachment reached an estimated 34 G (it was estimated because the accelerometer went out of range) before the engine's combustion chamber low-level pressure sensor commanded a shutdown. In the post-flight investigation, it was estimated that only one more cycle of amplitude growth could have been sustained without catastrophic failure. It was a wake-up call for NASA and the waiting pogo suppressor was installed for all subsequent Apollo missions, with the result that no further pogo was experienced.

After all their problems with pogo during the Gemini and Apollo missions, it wasn't surprising that NASA mandated that, for the Space Shuttle, there would be no pogo. The Space Shuttle, of course, was unique when compared to its predecessors. Structurally, it lacked the axial symmetry, and aerodynamic surfaces had been added. The main engines generated internal pressures three times as high as previous large engines, used a different approach to turbine drive, and used a full closed-loop thrust and a mixture ratio control system that was active into the pogo frequency range. Faced with these challenges, engine testing required the creation of entirely new approaches to the pogo problem, which resulted in a suppresser being designed as part of the engine. The suppression system developed for the Shuttle was an outstanding success and there was no pogo during the Space Shuttle era.

Having survived the acceleration dangers of launch and pogo problems, astronauts are faced with another gravity-related challenge, namely an absence of G, which causes a less violent but no less dangerous issue – muscle and bone deterioration. The term used to describe an absence of gravity in space is *microgravity* and it's the subject of the next chapter. For now, before moving onto the next G problem faced by astronauts, we just need to know that astronauts' bodies returning to Earth aren't as strong as when they left the planet, which is why re-entry can be a real workout, especially if that re-entry happens to be a ballistic one.

Re-entry

"I saw 8.2 Gs on the meter and it was pretty, pretty dramatic. Gravity's not really my friend right now and 8 Gs was especially not my friend. But it didn't last too long."

American astronaut, Peggy Whitson, describing her descent to Earth in a wayward Soyuz space capsule

On April 19th, 2008, Peggy Whitson, a Flight Engineer on board the ISS, was returning to Earth in the Russian Soyuz re-entry capsule. Accompanying her were Yi So-Yeon, South Korea's first astronaut, and Russian cosmonaut Yuri

Figure 7.8 Russian Soyuz approaching the International Space Station.
Image courtesy: NASA.

Malenchenko. In preparation for landing, their spacecraft, the Soyuz (Figure 7.8) TMA-11, undocked from the ISS and fired its braking engines. For a while, everything seemed to be proceeding according to plan. The next major re-entry event was the separation of the service module from the re-entry capsule. Unfortunately, this didn't happen. Fifty-five seconds after entering the upper atmosphere, the Soyuz's automated flight control system switched from aerodynamic to the ballistic mode of re-entry due to an attitude control error command (with the service module still attached to the crew capsule, the correct orientation wasn't possible).

It was the beginning of a rough ride.

About 100 seconds after the beginning of the re-entry, the service module finally tore off from the crew capsule. At this phase in the re-entry, the re-entry capsule and a loosely attached service module were flying entry hatch first but with a slight tilt, which exposed the belly of the re-entry capsule. As a result, a small gondola on the belly of the capsule that housed pitch attitude control thrusters burned through, becoming a source of smoke, which made its way inside the cabin and was reported by the crew. Fortunately, it happened late in the descent, when the loss of pressure was not as critical. As the smoke poured into the cabin, the Gs began to increase dramatically. During a nominal re-entry, the crew could have expected about 4 G. After the flight, Whitson reported seeing the G-meter pegged at more than 8 G. For someone who had spent 192 days in orbit, the Gs had a crushing effect on her deconditioned body.

After plummeting along its ballistic trajectory, the Russian TMA-11 craft "touched" (Whitson later described the landing as akin to a car crash) down on the Kazakh steppe but, due to the ballistic re-entry, the capsule was 400 km off course. It was the second time a Russian return vessel had landed off target.

Initially, Russian officials blamed the crew for making changes to the flight plan just before re-entry – a change that was not communicated to Mission Control, who assumed the original plan was going ahead. The result from this change was a steeper-than-normal angle on entering the atmosphere (a so-called ballistic re-entry), putting the capsule way off course. However, not all sources placed the blame on a lack of communication between capsule and Mission Control. In a controversial statement to reporters, Federal Space Agency chief, Anatoly Perminov, placed some of the blame on an old naval superstition that the female-dominated crew was bad luck and that Peggy Whitson and So-Yoen were responsible for the change of plan:

> "You know in Russia, there are certain bad omens about this sort of thing, but thank God that everything worked out successfully. Of course in the future, we will work somehow to ensure that the number of women will not surpass the number of men."
>
> Anatoly Perminov

The statement didn't endear Perminov to Whitson, but it caused a stir in the media, who fortunately were more interested in the Expedition Commander's assessment of events than what the Federal Space Chief had to say. Whitson, in a typical NASA understatement, described the re-entry as "challenging". She described going into a 17° per second spin to help stabilize the spacecraft during the ballistic portion of the descent and feeling her face pull back as the deceleration in the capsule built up to a punishing 8.5 G. Although the re-entry was rough, Whitson was quick to point out that astronauts trained for ballistic re-entries and it was considered a nominal form of re-entry, although they tried to avoid it whenever possible (it should be noted that, in the early days of space exploration, all re-entries were ballistic).

Surprisingly, given the seriousness of the event, NASA said it wasn't too worried about the capsule's off-course landing, despite a Russian space official telling the Russian news agency Interfax that the crew had been in serious danger. In fact, NASA associate administrator for space operations, William H. Gerstenmaier, downplayed any alarm, stating that he didn't see it as a major problem, although he also said it was clearly something that shouldn't have happened. House Science Committee Chairman, Bart Gordon, agreed, saying: "I'm obviously concerned anytime a human spaceflight mission doesn't go as planned. We need to get more information about what happened and why, as well as what will be done to keep it from happening again."

While the rest of the world tried to work out who was to blame for the Soyuz's botched landing, engineers tried to figure out what went wrong. In the aftermath of the Soyuz landing, the assembly of the next Soyuz vehicle was stopped until the problem was found. At the time, a failure of pyrotechnic devices in the PAO module separation system was suspected as the most probable culprit in the incident. However, the overload of the power supply system, which failed to deliver necessary electric current to the pyrotechnic devices, was also suggested. In this system, each connecting lock between the modules is attached via a dual pyrotechnic charge. According to unofficial reports, both charges on the suspect lock failed to fire. Also, telemetry reportedly showed a lack of electrical current on both pyrotechnic devices in one of the separation locks. For a while, attention was focused on the explosive bolts that possibly failed to fully separate the service module from the descent module as the craft began to enter the upper atmosphere. Another theory converged on the electromagnetic interference (EMI) hypothesis. EMI is thought to be caused by the flow of space plasma around the hull of the ISS, causing interference with the pyrotechnic bolts in the docked Soyuz vehicles. One effect could have been the hardening of the explosives igniter wire, meaning a higher electrical current was needed to trigger the small explosives.

While the TMA-11 incident was never satisfactorily resolved, the crew's harrowing ride once again put the dangers that astronauts face into the spotlight. And, as in so many challenges faced by spacefarers, G featured prominently.

Returning from space

INT. COCKPIT. DAY
With alarms sounding, the four astronauts are slammed violently forward, against their restraints.

                    WOODY
        What the --

                    MCCONNELL
        Shut down engines!

He and Woody both reach out, straining against the incredible G-forces, and manage to grab a red emergency lever, yanking it down hard.

EXT. MARS RECOVERY. DAY

Too late! IN SLOW MOTION, we see a terrifyingly violent chain reaction. The fuel tanks themselves explode, one after another. The supporting metal struts are vaporized. The solar panels are snapped off. Two of the huge engine bells are smashed sideways, out of alignment, while the third, trailing pieces of the cowling, goes hurtling off, like a flaming cannon ball. The entire aft section of the ship, including much of the lower Hab, instantly becomes a shredded, charred tangle of metal, and even worse, the explosion causes what's left of the ship -- mainly the EVA chamber and cockpit to tumble end over end, cartwheeling down towards Mars.

INT. COCKPIT. DAY

In the windows, Mars goes crazily in and out of view. The astronauts, flung this way, then that, are all fighting against unconsciousness.

> WOODY
> Engines negative! No response! I've got no attitude control!

> MCCONNELL
> Manual separation! Blow the bolts!

> WOODY
> Negative! The CM doesn't have enough thrust to correct this rotation!

> PHIL
> We're too steep! Falling into the atmosphere...!

EXT. MARS RECOVERY. DAY

From further away, the charred remainder of the ship can be seen spinning down towards Mars, which now fills the screen, looming as large as Earth, when seen from the space shuttle. The ship's motion is mercifully slowing as it hits the outer atmosphere, but just as clearly this steep, unplanned angle of entry dooms it.

> PHIL
> Christ, at this angle we'll burn up!

INT. COCKPIT. DAY

ON A COMPUTER.

Their ANGLE OF ENTRY is shown -- much too direct -- with an indicated swerve into blinking red disaster. Warnings flash: CRITICAL ENTRY! PULL OUT!

> Excerpt from the screenplay, *Mission to Mars*, by Jim Thomas, John Thomas, and Graham Yost, story by Lowell Cannon. Courtesy: IMDB

The above excerpt is typical Hollywood but it alludes to the fact that any spacecraft re-entering an atmosphere, whether it be Martian or terrestrial, must do so within a very small range of angles if it is to reach the surface safely. The upper and lower limits of this re-entry region are determined by a combination of three factors: the spacecraft's trajectory, its rate of deceleration, and aerodynamic heating. The narrow region dictated by these parameters is known as the *entry corridor*. The trajectory a spacecraft follows when returning to Earth depends in part on the type of orbit the vehicle traveled and the orbital path is significant because it determines how fast the vehicle is traveling when it first encounters the atmosphere. One type of orbit is the circular orbit that approximates the path most spacecraft follow while orbiting Earth. Vehicles traveling in a circular orbit typically circle Earth at speeds of 27,360–28,970 km/hr (nearly 30 times the speed of sound, or about 7.5 km/second) and will re-enter the outer fringes of the atmosphere at those speeds. The first stage in making a safe landing is for a returning spacecraft to lose nearly all of that orbital speed. The operation is basically a reversal of the launch phase, and this means the returning vehicle must "sink" as much kinetic energy as the propulsion systems generated between lift-off and orbit. Theoretically speaking, there are three fundamentally different methods by which today's spacecraft can do this: powered deceleration, mass shedding, and energy dissipation.

Powered deceleration can be achieved by rocket thrust opposed to the direction of motion. The Space Shuttle used to do this at the very beginning of its re-entry, to trim the speed and initiate the descent from orbit. The equivalence of re-entry and launch energy means that it's not a practical proposition for the whole descent, because using this method would more than double the fuel load required at launch and would require the vehicle to carry half of it while in space.

Mass shedding can be conceptually exemplified by a pilot ejecting from a damaged aircraft, since most of the system's energy remains invested in the part that carries on and crashes. If a spacecraft is dispensable, astronauts can transfer into a small escape vehicle which can then be jettisoned from the main craft. The re-entry phase of space programs such as Apollo made use of mass shedding.

Energy dissipation refers to kinetic energy being progressively converted to another form, such as heat, as the descent proceeds. It's the principal method for re-entry braking in all mankind's space programs to date and it's a technique that requires an enormous amount of energy. It also happens to be the means by which the Soyuz spacecraft – a 1960s-era vehicle – returns its human cargo to Earth, which brings us back to the subject of ballistic re-entry.

Types of re-entry

In the earlier pre-Shuttle era of manned spaceflight, descent through the atmosphere was always ballistic and it's a method that remains the simplest and probably the safest way to return to Earth from space. Its disadvantage is that it only works for a residual capsule of small size and constrained geometry, and a great deal of equipment has to be ejected to strip down to the ballistic re-entry configuration, which means it's a costly method (the Soyuz capsules can only be used once). Apollo's command module can also be considered to be an example of a ballistic re-entry vehicle. It weighed about 6 tonnes and was basically a steel cone just large enough to accommodate three men. The ablative heat shield accounted for nearly a quarter of the initial weight of the module. Ultimately, all practical re-entry vehicles devised to date dissipate their kinetic energy by converting it to heat. The Shuttle was no exception but, since it weighed about 100 tonnes, it had to lose nearly 20 times more energy than the Apollo command module, which of course meant it had to generate and dissipate 20 times as much heat. If it had had to do that in the short time it would take to fall from orbit along a ballistic re-entry trajectory like the Soyuz, it would have been incinerated – even if constructed using the most advanced materials available today – so a ballistic re-entry isn't practical for a large reusable spacecraft.

The alternative to ballistic re-entry is to descend at a lower rate than free fall. An acceptable rate of descent is one at which the rate of heating is reduced to a manageable level. Since the Space Shuttle couldn't use powered rockets to slow its fall, it had to fly, to generate upward force (termed "lift") as a result of its forward movement that kept it airborne. Since the Shuttle was unpowered (except for thrusters used to orient the vehicle), it was technically a glider, and a poor one at that, being much too heavy and squat to generate significant low-speed lift, which was why it was nicknamed the "flying brick".

About 70 minutes before touchdown, the actions that committed the Shuttle to re-entry were taken. Until this point, it was in a stable orbit in a nose-first, belly-up attitude at an altitude of 150 km with a velocity of a little over 29,000 km/hr. At this stage, the Shuttle wasn't flying, because there was no substantial atmosphere for flight surfaces to bear on at that altitude. At approximately 35 minutes before touchdown, 8,000 km from the landing site and at an altitude of around 120 km, the Shuttle entered a discernable atmosphere. At this point, the astronauts began to feel the G but nothing like the Gs experienced (suffered) by the TMA-11 Soyuz crew. In fact, G-loads on board the Shuttle didn't exceed 1 G until 600 seconds into the re-entry profile, which was at 73,000-m altitude: the start point of the true deceleration in the thicker atmosphere. From this point, the angle of attack was critical and was maintained at 40° by automatic thruster trims. At an altitude of 85 km, the flight surfaces of the orbiter became usable. Under automatic control, a series of four S-bend turns was now performed, with the craft banking through 80° at the fullest extent of the roll, with the effect that the Shuttle gradually bled off speed. Ten minutes before touchdown, the Shuttle was still 40 km up and traveling at nearly 10,000 km/hr (Mach 8). Drag on the craft increased greatly and deceleration took place more rapidly, resulting in higher Gs being experienced by the crew, although the 1.7-G peak was nothing like the 8-plus G Whitson's crew was subjected to. At about 40 km from the runway, the pilot began to adjust the heading during the aggressive descent. Ninety seconds from touchdown, the altitude was still around 4,000 m, and the Shuttle lost altitude about 20 times quicker than a commercial airliner. At 600 m and 30 seconds from landing, the pilot pulled up the nose once more and deployed the landing gear. The speed at touchdown was typically 350 km/hr. As the speed fell to about 150 km/hr, the nose-wheel touched down and full braking was employed. The parachute was jettisoned just before the craft rolled to a complete stop, about 2 km from its point of touchdown.

Thanks to its gliding ability that allowed it to progressively shed speed, the Shuttle imposed only mild G-forces on returning astronauts, but that isn't the case for vehicles returning from beyond Earth orbit. Instead of traveling in circular paths, these craft follow parabolic or hyperbolic orbits that result in much higher speeds upon returning to Earth. Apollo capsules, for example, re-entered the atmosphere at speeds of nearly 40,000 km/hr and followed the uncomfortable ballistic re-entry profile. In this type of re-entry, the vehicle generates very little aerodynamic lift and instead plunges into the atmosphere and falls through it under the influence of gravity and drag. This drag force slows the vehicle so that parachutes can be deployed for a soft touchdown. The landing point is predetermined by conditions when the vehicle first enters the atmosphere, and the pilot has no control over the capsule's trajectory or its landing point once he leaves orbit and begins the ballistic plunge. That's assuming everything goes to plan, since there are limits to how steeply a vehicle can re-enter the atmosphere and land safely. If the entry angle is too high, the craft will decelerate too quickly and the G-forces will crush the occupants. If the entry angle is too shallow, the vehicle will generate too little drag and not slow down enough to follow a trajectory down to the surface.

Re-entry on Mars

The challenge of a ballistic re-entry and associated G-loads is even more of a problem when it comes to devising a way of landing humans on Mars. First of all, there's too much atmosphere on Mars to land heavy vehicles like we do on the Moon, using propulsive technology, and there's too little atmosphere to land like we do on Earth. From a re-entry perspective, Mars is in a gray zone. But what about airbags, parachutes, or thrusters that have been used on the previous successful robotic Mars missions, or using a lifting body vehicle similar to the Space Shuttle? Well, these systems work when we're talking about landing small payloads (about 1 tonne) but nobody has figured out how to safely get large masses down to the surface of Mars. The problem of landing a large (manned) payload on the surface has been dubbed the Supersonic Transition Problem (STP), since there is a velocity-altitude gap below Mach 5. The gap is between the delivery capability of large entry systems at Mars and the capability of super and subsonic decelerator technologies to get below the speed of sound. Put simply, with current capabilities, a large, heavy vehicle, plummeting through the thin Martian atmosphere only has about 90 seconds to slow down (Figure 7.9) from Mach 5 to under Mach 1, change and re-orient itself from being a spacecraft to a lander, and then deploy parachutes to decelerate some more, before having to use thrusters to finally touch down.

When the STP is presented to people, the most common suggestion is to use airbags. After all, airbags were successful in landing the Pathfinder rover and the two Mars Exploration Rovers. But an airbag landing subjects the payload to forces of between 10 and 20 G, which is fine for robots but not so good for humans. So,

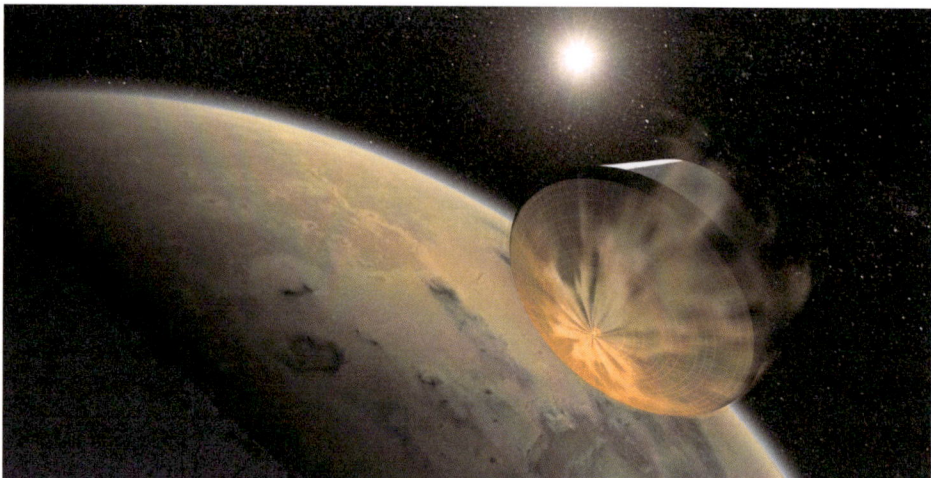

Figure 7.9 Martian re-entry. Image courtesy: NASA.

Figure 7.10 The Sky Crane concept will be used to slow down the Mars Science Laboratory (MSL) as it enters the Martian atmosphere. The rockets of the aeroshell will fire to steer the capsule towards the desired angle. As soon as the capsule slows down, the heat shield will eject, leaving the rover exposed inside the aeroshell, attached to the floating crane mechanism. The floating crane's rockets will fire to further slow the descent. The top part of the aeroshell will detach, leaving the Sky Crane alone holding the MSL rover, descending towards the planet's surface. A few hundred meters above the terrain, the floating Sky Crane will start lowering the rover down using a trio of bridles and one umbilical cord until it touches down. At that time, the Sky Crane will detach from the rover and fly away. Image courtesy: NASA.

just how *do* you land humans on Mars without crushing them with G-forces? Well, the latest landing system is the Sky Crane (Figure 7.10), which uses the combination of a rocket-guided entry with a heat shield, a parachute, and thrusters to slow the vehicle even more. Once the vehicle's speed is slow enough, a crane-like system lowers the vehicle on a cable for a soft landing. It's a system that might be scaled for larger payloads, but engineers aren't sure whether it will be able to land humans safely on Mars.

So, what about parachutes? Well, Mars's atmosphere is only 1% as dense as Earth's, which means Mars's atmosphere at its thickest is equivalent to Earth's atmosphere at about 35 km above the surface. Parachutes – even supersonic ones – can only be opened at speeds of less than Mach 2 and there's just no way a heavy spacecraft heading towards Mars could ever be slowed down that much using just a heat shield. That's not to say parachutes can't be used, but they would have to be big. Really big. Supersonic parachute experts reckon that to slow a manned vehicle on Mars and reach the surface at reasonable speeds would require a parachute 100 m in diameter. The problem would be opening the chute – there's no way to make a 100-m parachute that can be opened safely supersonically (think about the G-forces during the shock of opening), not to mention the time it takes to inflate something that large. You'd be on the ground before it was fully inflated. It would not be a good outcome.

Okay, no parachutes, so what about a higher lift vehicle like the Space Shuttle? Well, on Mars, when you use a very high lift-to-weight-to-drag ratio to get good deceleration and use the lift properly, the vehicle would have to be very low in the atmosphere. In fact, the vehicle would still be speeding along at Mach 2 or 3 fairly close to the ground and would have to slow to under Mach 2 before opening a parachute. By this time, it would be too late.

Well, if parachutes are no good and lift vehicles won't cut it, what about a good old-fashioned heat shield? It's not inconceivable that engineers can design larger, lighter heat shields, but the problem is that, using current technology, the heat shield diameter for a human-capable spacecraft would be so big that it couldn't be launched from Earth. Thrusters could be used, but only at great cost because, for every kilogram of payload in orbit, it takes twice as much fuel to get to the surface of Mars as to the Moon. Carrying along thruster propellant would entail perhaps over six times the payload mass in fuel to get human-sized payloads to the surface, all of which would have to be brought along from Earth. That just isn't an option but, even if somehow you *could* transport all that fuel, using current thruster technology in Mars's atmosphere poses aerodynamic problems because rocket plumes are notoriously unstable, dynamic, chaotic systems. Even though the Martian atmospheric density is very low, because the re-entry velocity is so high, these forces might cause such extreme stress on the vehicle that it might be torn apart by all the shaking and twisting.

Using thrusters in combination with a heat shield and parachute isn't without problems either. Assuming the vehicle could use some technique to slow to under Mach 1, using propulsion only in the final stages of descent to gradually adjust the vehicle's trajectory would enable the vehicle to arrive very precisely at

Figure 7.11 A Hypercone is a mechanism for atmospheric re-entry deceleration. It is an inflatable structure combining characteristics of both heat shields and parachutes. The Hypercone is intended to supplement other deceleration mechanisms, bridging a gap in capability between conventional heat shields and conventional parachutes or landing rockets. A Hypercone consists of a large donut-shaped balloon that supports a cone-shaped sheet of heat-resistant fabric 30–40 m in diameter, with the capsule located at the point of the cone. The balloon is rapidly inflated to expand the cone to full size and the resulting drag slows the capsule to a velocity at which point other landing mechanisms can finish the job of bringing it to a soft landing. Image courtesy: NASA.

the desired landing site. In this scenario, the crew would be firing thrusters less than 1 km above the ground. From a deceleration perspective, it would probably be the safest re-entry method but, with just 1,000 m of altitude to play with, even this combination wouldn't give the crew much latitude to fly out any landing uncertainties in the event of mis-targeting.

Perhaps the best hope on the horizon for making a safe landing on Mars is a supersonic decelerator called a Hypercone (Figure 7.11). Imagine a huge donut with a skin across its surface that girdles the vehicle and inflates very quickly with gas rockets (like air bags) to create a conical shape. This would inflate about 10 km above the ground while the vehicle is traveling at Mach 4 or 5. The Hypercone would act as an aerodynamic anchor to slow the vehicle to Mach 1. Vertigo, a company that designs these sorts of things, has been doing extensive analysis of the Hypercone, including sizing and mass estimates for landers from 4 to 60 metric tonnes.

Vertigo is currently competing for funding from NASA for further research, as the next step – deployment in a supersonic wind tunnel – is quite expensive. The structure would need to be about 30–40 m in diameter. The problem here is that large, flexible structures are notoriously difficult to control. Also, there are several unknowns in developing and using a Hypercone. Some engineers reckon that if the Hypercone can get the vehicle under Mach 1, then subsonic parachutes could be used, much like those employed by Apollo. But, it takes time for the parachutes to inflate and there would only be a matter of seconds of use, allowing time to shed the parachutes before converting to a propulsive system. And the thrusters would be absolutely necessary because the vehicle would be falling 10 times faster than through Earth's atmosphere because the density of Mars's atmosphere is 100 times lower than Earth's. Put simply, this means the vehicle couldn't just land with parachutes and touch the ground because the impact would break the astronauts' bones, if not the hardware. Also, spending time in space (microgravity) results in deconditioning, which means astronauts preparing to land on Mars are in a weakened state, so, while exposure to 3 G poses no problems for astronauts during take-off, even this moderate level of G-force could be problematic during landing.

8 Microgravity

The zero-G challenges of being an astronaut

Many people mistakenly think that gravity doesn't exist in space. However, typical orbital altitudes for astronauts vary between 200 and 550 km above Earth's surface and the gravitational field is still quite strong at these altitudes. In fact, Earth's gravitational field at about 400 km above the surface is 88.8% of its strength at the surface – it's the reason why the International Space Station (ISS) is kept in orbit around Earth. To understand how the ISS remains in orbit, we have to revisit the nature of gravity. As first described by Sir Isaac Newton more than 300 years ago, gravity is the attraction between two masses and the acceleration of an object towards the ground caused by gravity alone, near the surface of Earth, is called "normal gravity", or 1 g (this acceleration is equal to 9.8 m/second^2). Now, if you drop an object on Earth, it falls at 1 g and, if an astronaut on the ISS drops an object, it falls too – it just doesn't look like it's falling. That's because the object, the astronaut, and the ISS are all falling together. But they're not falling *towards* Earth; they're falling *around* it. That's because they're all falling at the same rate and objects inside the ISS appear to float in a state that scientists call microgravity.

Fortunately, for those wanting to experience the sensation of floating around, you don't have to qualify as an astronaut or be rich enough to buy a ticket as a space tourist because microgravity can be created a little closer to home. To understand how microgravity is created, it's necessary to revisit basic physics. First, we need to understand that microgravity occurs whenever an object is in free fall. In other words, the object falls faster and faster, its acceleration caused by gravity. As soon as you drop something, it is in a state of free fall and the same principle applies if you throw something, since it immediately starts falling towards Earth. But how does something fall *around* Earth?

Newton came up with an idea to demonstrate the concept. Imagine placing a cannon on top of a very tall mountain. Once fired, a cannonball falls to Earth and the greater the speed, the farther the cannonball will travel before landing. If fired with the proper speed, the cannonball would achieve a state of continuous free fall around Earth, which is termed *orbit*. It's the same principle that applies to the ISS. While objects inside the ISS appear to be floating and motionless, they're actually traveling at the same orbital speed as the space station: 28,000 km/hr.

Parabolic flight

NASA uses a variety of facilities to create microgravity conditions. For example, the agency has a drop tower called the Zero Gravity Research Facility, which is a large shaft more than 150 m deep that allows test packages to free fall in a vacuum for just over five seconds. In this state of free fall, weightlessness (at or near microgravity) can be achieved. While the drop tower is a good way of testing equipment, it's no good for training astronauts, which is why NASA and other space agencies rely on specially adapted aircraft to fly in parabolic arcs to create microgravity for tests and simulations that last 20–25 seconds. For those

Figure 8.1 This is ESA's current zero-G aircraft, which replaced the dated Caravelle in the 1990s. The aircraft gives its occupants the sensation of weightlessness by following a parabolic flight path relative to the center of Earth. While following this path, the aircraft and its payload are in free fall at certain points of its flight path. Initially, the aircraft climbs with a pitch angle of 45°. The sensation of weightlessness is achieved by reducing thrust and lowering the nose to maintain a zero-lift angle of attack. Weightlessness begins while ascending and lasts all the way "up-and-over the hump", until the craft reaches a declined angle of 30°. At this point, the craft is pointed downward at high speed, and must begin to pull back into the nose-up attitude to repeat the maneuver. The forces are then roughly twice that of gravity on the way down, at the bottom, and up again. This lasts all the way until the aircraft is again halfway up its upward trajectory, and the pilot again initiates the zero-G flight path. Image courtesy: ESA.

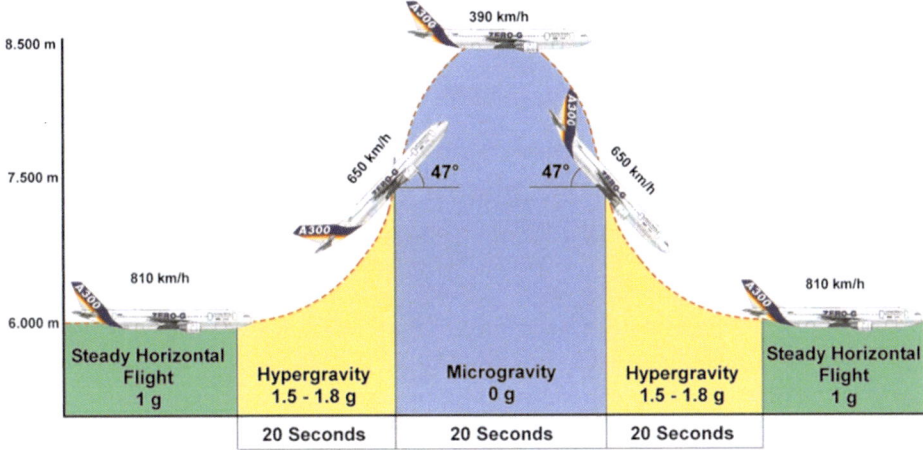

Figure 8.2 Parabolic flight explained. Image courtesy: ESA.

who are interested in experiencing the thrills of microgravity and who have $5,000 to spare, there is Zero-G Corporation's G-Force One, a modified Boeing 727-200. If you don't have $5,000 lying around, you can always do what I did and volunteer as a research subject. In 1995, one of my Ph.D. supervisor's friends, Dr Paul Enck, received a grant from the European Space Agency (ESA) to conduct testing on board the Caravelle that ESA (Figure 8.1) used to conduct parabolic flights at the time. Back in those days, there was no Zero-G Corporation, so this was a rare opportunity to experience microgravity. In the 1990s, a typical ESA flight campaign consisted of three flights of 30 parabolas per flight. On each parabola, there was a period of increased gravity (1.8–2 G) that lasted for about 20 seconds immediately prior to and following the 20-second period of reduced gravity (Figure 8.2). It was a lot of fun, as I was lucky enough to fly 30 parabolas for a cumulative total of 13 minutes of microgravity.

> "It took only one flight in the jet to understand why it was named the Vomit Comet. The plane was a barf factory. Just climbing aboard, the nose would detect a faint odor of bile. Like cigarette smoke that cannot be removed from the drapes of a two-pack-a-day addict, the smell of stomach fluid had permeated the very aluminum structure of the machine."
>
> Astronaut, Mike Mullane, describing his experience
> on board NASA's modified Boeing 7070 aircraft.
> Excerpt from Mike Mullane's excellent memoir, *Riding Rockets* [1]

For investigators, parabolic flights (Figure 8.3) offer a number of advantages. First, the lead time is short, with only a few months between research proposal and flight; second, if you happen to be a scientist flying an experiment with ESA, the space agency covers the cost of the flight; and third, investigators on board

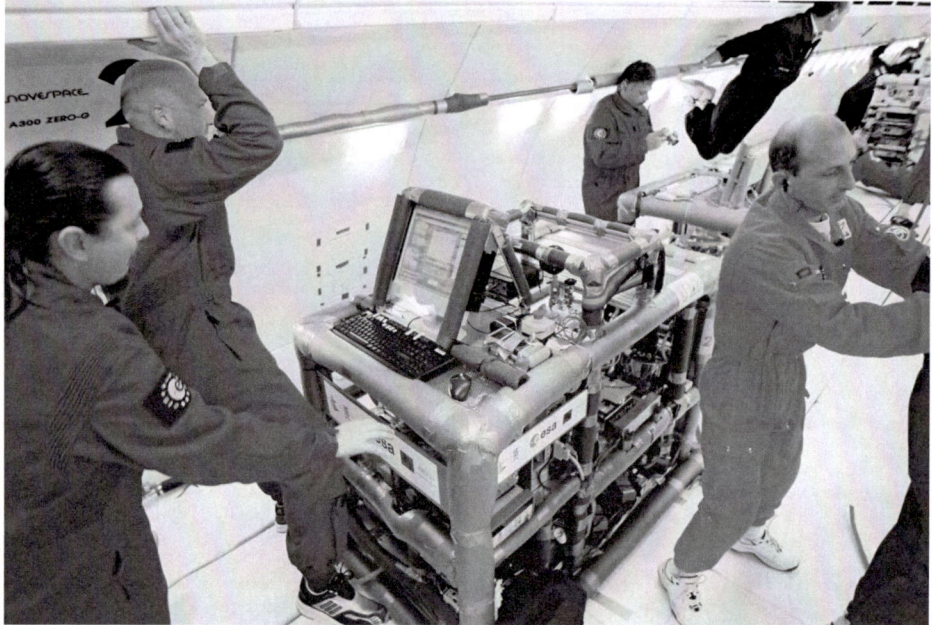

Figure 8.3 Scientists conduct research on board a parabolic flight. Image courtesy: ESA.

the aircraft can interact with their experiments between, and even during, parabola – as long as the scientists aren't prone to motion sickness (see Mike Mullane's quote).

Today, the main use of parabolic flights is still to conduct science, to test instrumentation prior to use in space, and to validate operational and experimental procedures. It's also used to train astronauts for future spaceflights and, as every astronaut knows, the key to any successful mission is practice, practice, practice, which brings us to another training aid: the Neutral Buoyancy Laboratory (NBL).

Neutral Buoyancy Laboratory

While parabolic flight indoctrinates astronauts to the world of microgravity, the usefulness of zero-G as a training aid is limited, since many mission activities such as extravehicular activities (EVAs) last several hours. But how do you simulate such an environment on Earth? After all, astronauts conducting EVAs work in a vacuum outside Earth's atmosphere and must be protected by bulky environmental suits. They must also learn to function within these suits to perform useful tasks and accomplishing these tasks requires special tools and a lot of practice. To do this, astronauts spend time in the NBL (Figure 8.4).

The NBL is an astronaut-training facility maintained by and located at the

Figure 8.4 The Neutral Buoyancy Laboratory (NBL), which contains 23.5 million liters of water, is an astronaut training facility maintained by and located at the Sonny Carter Training Facility in NASA's Johnson Space Center in Houston. Image courtesy: NASA.

Sonny Carter Training Facility at NASA's Johnson Space Center (JSC) in Houston. At 62 m long, 31 m wide, and 12 m deep, it's the largest pool in the world and is the training location for astronauts who need to practice EVA tasks. Very simply, the principle of the NBL is to simulate the weightless environment of space. First, the suited astronauts and/or equipment are lowered into the pool using an overhead crane (Figure 8.5). Then, the astronauts are weighted in the water by support divers so they experience no buoyant force and no rotational moment about their center of mass.

The pool itself appears no different from any other swimming pool: it's filled with water and reeks of chlorine. The surroundings, however, rapidly inform the visitor that this is no ordinary swimming pool. Above the pool are two pneumatic cranes, capable of lifting full-scale payload mockups, and on one side of the pool is a storage area for a vast assortment of these full-scale models. Different training missions require different mockups and the cranes move the relevant mockups into and out of the pool. In addition, there are two monitoring stations: one that monitors the physiological status of the two astronauts in the pool during a training mission and a second series of monitors that continuously evaluates the environmental suits the astronauts use. Next to this monitoring station is all the plumbing necessary to provide breathing air and cooling water to the astronauts' suits. Lastly, there is a complete diving locker equipped with compressor and storage cylinders to support a team of more than 30 divers.

The NBL dive support team aren't your typical sport divers, since they must be retrained to much higher standards to operate within NBL working guidelines and each must be qualified in astronaut in-pool rescue. This involves lifting a diver in a totally flooded environmental spacesuit from the bottom of the pool to a position where life-support assistance is possible – an exercise that must be successfully completed in less than one minute.

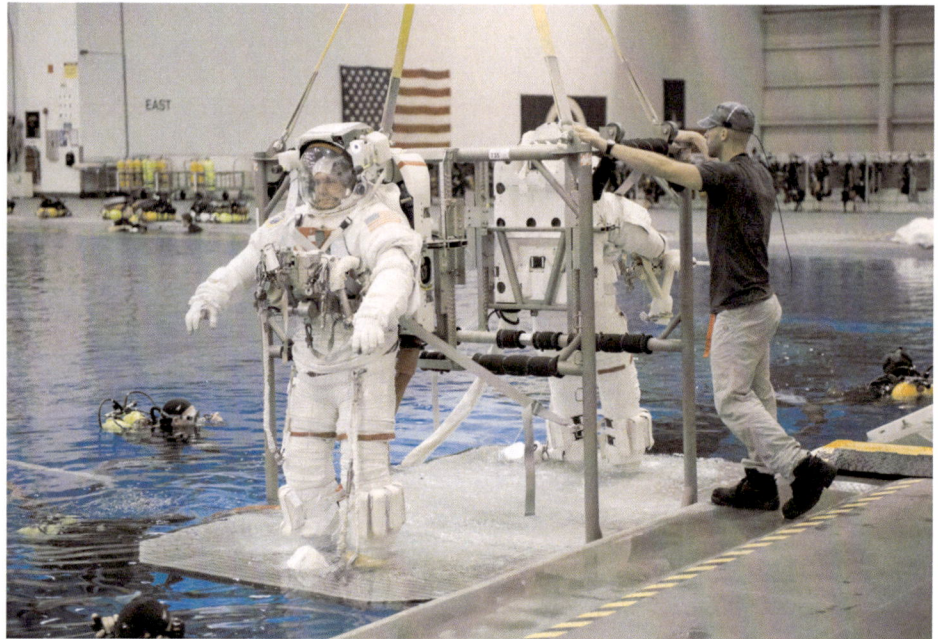

Figure 8.5 An astronaut is lowered into the NBL. Image courtesy: NASA.

Before a training session begins, the divers inspect the training mockups and ensure that all necessary materials have been assembled and are available to the astronauts. During a training mission, each astronaut is assigned two divers. The training mission begins by ensuring that the astronaut and all his/her limbs are neutrally buoyant. This involves a rather elaborate system of weights that are placed at various locations on the environmental protection suit so the astronaut can be truly neutrally buoyant. Of course, each astronaut is physically different, so each astronaut's buoyancy must be individually tuned. The goal is that no matter what orientation the astronaut assumes in the pool, neither gravity nor suit buoyancy will control movement. Throughout the training mission, the divers, as well as the topside crew, monitor the astronaut. After the training mission is over, the divers reconfigure the pool and ensure all training mockups are ready for the next training mission.

Long-term effects of microgravity

Following the advent of space stations such as the Mir and the ISS that could be inhabited for long periods of time, exposure to microgravity has proven to exert some deleterious effects on astronauts. Perhaps one of the most common problems experienced by astronauts during the initial hours of microgravity is known as space adaptation syndrome (SAS), commonly referred to as space sickness. Some parabolic flight participants experience a similar syndrome, which is why the aircraft I flew on had a couple of assistants who patrolled through the cabin between each parabola handing out sick bags. Symptoms of SAS include nausea and vomiting, vertigo, headaches, lethargy, and overall malaise. The first case of SAS was reported by cosmonaut Gherman Titov in 1961 and, since then, more than three-quarters of first-time astronauts experience symptoms. The duration of SAS varies but it generally lasts for a day or two, and no longer than three days. NASA jokingly measures SAS using the "Garn scale", named after US Senator Jake Garn, whose SAS symptoms during Space Shuttle flight STS-51-D were the worst on record.

Space motion sickness is caused by changes in gravitational forces, which affect spatial orientation. The transition to microgravity that occurs during spaceflight influences our spatial orientation and requires adaptation by our balance system (Figure 8.6). As long as this adaptation is incomplete, astronauts suffer typical symptoms of motion sickness. There are motion-sickness medications that astronauts can take but these tend to make the crew drowsy, so they are rarely used, although transdermal dimenhydrinate is used as a backup measure. While SAS is a significant problem, the syndrome is a short-term effect that only lasts a couple of days. More serious are the long-term effects such as muscle atrophy (loss) and deterioration of the skeleton.

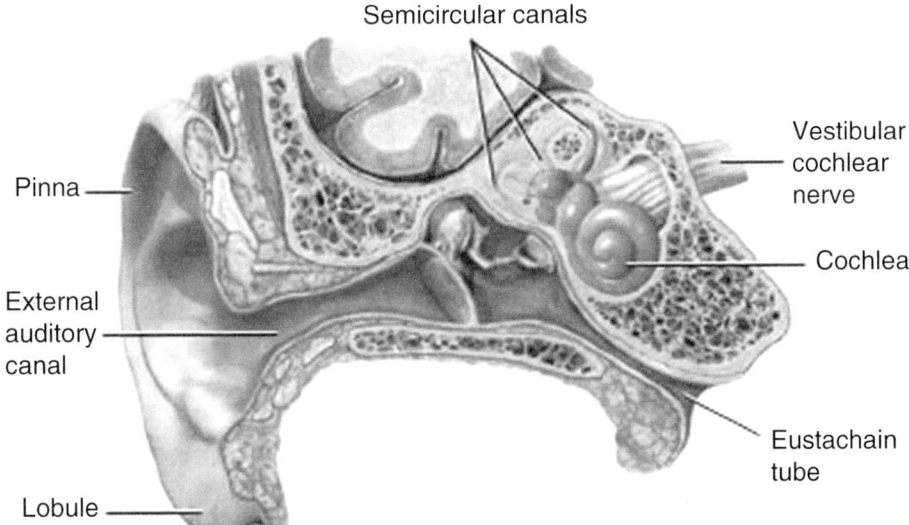

Semicircular canals

Pinna

External auditory canal

Lobule

Vestibular cochlear nerve

Cochlea

Eustachain tube

Figure 8.6 The vestibular system is the sensory system that provides information about movement and sense of balance. As our movements consist of rotations and translations, the vestibular system comprises two components: the semicircular canal system, which indicates rotational movements, and the otoliths, which indicate linear accelerations. Image courtesy: A.D.A.M.

Bone demineralization

Bone demineralization begins immediately on arrival in space. During the first few days in microgravity, a 60–70% increase in the amount of calcium excreted by the body is observed. The depletion is rapid and continuous, resulting in losses of bone mineral, changes in bone architecture, and alterations in skeletal mass that result in a condition similar to osteoporosis. This microgravity-induced loss of bone mineral density (BMD) has been documented primarily in the weight-bearing components of the skeletal system such as the lumbar vertebrae, femoral neck, trochanter, tibia, and calcaneus. Research on board the ISS has shown that astronauts may lose between 1 and 2% of their BMD per month – a rate almost five times that of women with postmenopausal osteoporosis! This is bad news for astronauts working on board the ISS during their six-month increments, but worse for those who eventually make the two-to-three-year round trip to Mars. These explorers could lose 20% of their BMD, equating to a 40% loss in bone strength.

Although the reduced (one-third) gravity of Mars will lessen the effect of bone demineralization, the sheer magnitude of bone loss experienced en route to the destination will mean astronauts will still be highly susceptible to the risk of fracture. Furthermore, in the event of a crewmember suffering a fracture, healing would be inhibited due to the reduced Martian gravitational field. Then there are the problems of astronauts returning from Mars never fully recovering their bone mass and the related health hazards such as toxic accumulations of excess mineral in the kidneys. It's a major concern among mission planners, which is why scientists have developed countermeasures. But, before we discuss these, it's useful to understand the physiological processes that occur in the skeletal system during exposure to microgravity.

The absence of gravitational load removes not only the direct compressive forces on the long bones and spine, but also the indirect loading on these bones from the pull of muscles on the various bone structures to which they are attached. Invariably, the unloading of the skeleton leads to osteoporosis (Figure 8.7), weakening of the bones, and delayed healing of fractures.

Bone is composed of mineral and organic components. Collagen, the most abundant protein in bone, is synthesized primarily by osteoblasts (bone cells responsible for removing bone tissue) and forms a framework upon which mineralization is superimposed. Adding to this process are various matrix proteins that have cell-recruitment functions in remodeling bone. At present, there is little information concerning the influence of microgravity on the biophysical functions of these matrix proteins. Compounding the issue is the possibility that an impaired mineralization process may occur during spaceflight. Bone demineralization is a complex and dynamic sequence of events involving mineral deposition regulated by cells responsible for aligning calcium phosphate crystals and depositing them within the collagen structure. Evidence from spaceflight indicates these minerals, when formed in microgravity, have a decreased crystal size and are configured imperfectly.

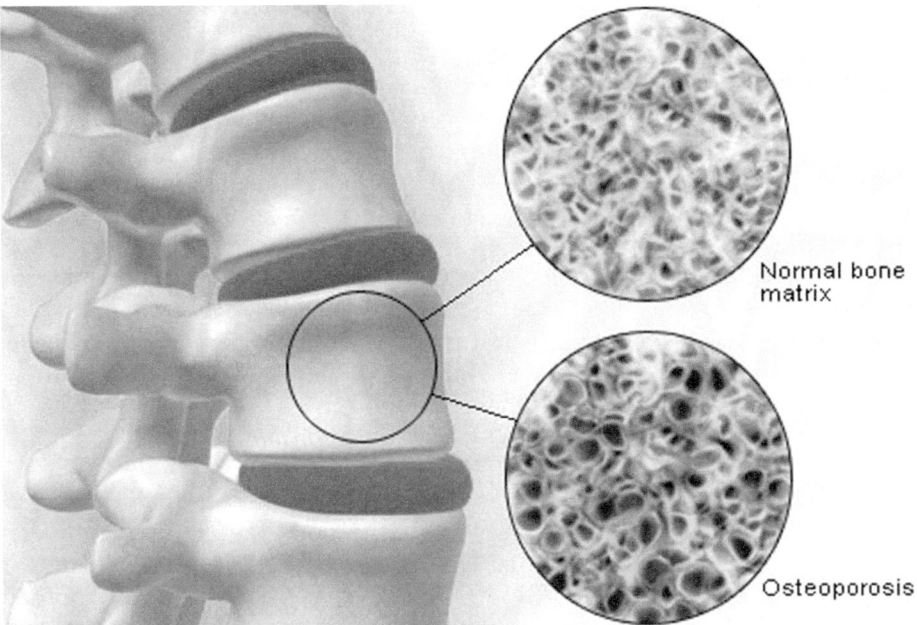

Normal bone
matrix

Osteoporosis

Figure 8.7 Bedridden patients and astronauts share something in common: progressive bone loss. Immobile patients lose bone density because they don't exercise muscles that would otherwise build skeletal strength through motion. Astronauts also face long periods of immobility, in addition to zero gravity, which negatively affects bone cell function. Astronauts in space can lose up to 2% of bone mass per month, which is several times more than is lost by patients with osteoporosis. Bone cell formation depends largely on the effects of weight, through both gravity and exercise. When weight is suppressed, bones undergo a process of demineralization accompanied by a loss of calcium to the blood. Image courtesy: NASA.

Also important to the understanding and prevention of bone loss during microgravity are the processes of repair and remodeling. The remodeling process is governed by osteoblasts and osteoclasts (bone cells responsible for bone repair), although the control mechanism has not been identified. It is known that, in space, the astronaut's skeleton undergoes a fast rate of resorption due to the unloading of mechanical stresses and weights. As a result of the skeleton no longer having to bear the astronaut's full weight, the body signals to the osteoclasts to resorb bone at a fast rate and thereby begins to rid itself of what it believes is unnecessary bone. This process occurs in tandem with the rate of bone formation, which is negatively affected by microgravity, resulting in a slowing of the action of osteoclasts and a reduction in the amount of calcium absorption. For example, on Earth, bone absorbs 40–50% of the calcium intake, whereas only 20–25% is absorbed in space.

Another component of the bone-loss mystery is the process of bone homeostasis. Bone tissue is constantly recycled and renewed to maintain homeostasis – a process of bone-remodeling and repair resulting in approximately 500 mg of calcium entering or leaving the bone each day. This remodeling occurs selectively in a process of reabsorbing or depositing bone tissue determined by the mechanical or gravitational stresses acting on the bone. Together, the osteoclast and osteoblast cells remodel bone tissue continuously in a process controlled by hormonal and mechanical feedback. Unfortunately, removal of gravitational stress results in a disruption to both of these feedback processes, resulting in bone atrophy – a situation compounded by the effect of the blood supply.

We know that all the physiological processes in bone depend on an optimal bone blood supply but, to understand how blood supply can cause bone atrophy, we must also be familiar with the effect of the absence of gravity on the circulatory system. Gravity affects an organism *hydrostatically* so, when an astronaut on Earth is in an upright position, the proportion of fluid volume in their lower half is greater than in their upper half. But, once the force of gravity is removed, the hydrostatic forces exerted on bodily fluid are completely neutralized and blood is distributed evenly throughout the body. This means the body detects less blood in the extremities such as the legs and its response to this unnatural blood redistribution is to pump more blood through the heart. Unfortunately, the increase in blood circulation leads to *accelerated* demineralization because increased blood flow results in an increased blood velocity through bone that increases the rate of calcium absorption into the blood supply. It really is a "lose–lose" situation.

Obviously, with all these problems associated with bone loss, it will be important for Mars-bound crews to monitor their rate of bone loss and BMD. Some of the current methods of measuring BMD include ultrasound, computed tomography, magnetic resonance imaging (MRI), and Dual-Energy Absorptiometry (DXA). Of these, perhaps the most accurate is DXA (Figure 8.8), a system in which two low-dose X-ray beams of different energies are used to scan regions of the body suspected of bone loss.

The reason two different X-ray energies are used is to distinguish between bone and muscle, since each tissue absorbs differently. Although the results from a DXA scan provide a reasonably accurate determination of BMD, one of the drawbacks of the system is its inability to distinguish between compact and cancellous bone, making it almost impossible to reconstruct an engineering model of the bone to perform the necessary stress-loading simulations. Since it's necessary to determine the specific location of bone loss to accurately assess fracture risk, a more sensitive means of assessing BMD is required. For interplanetary missions, this equipment also needs to be flight-qualifiable. Such a system might be a scaled-down version of the Advanced Multiple-Projection Dual-energy X-ray Absorptiometry (AMPDXA) system, which allows a much-higher-resolution and precision image to be produced. Because the system uses multiple images acquired at different angles, it is possible to determine precise BMD and bone geometry images that may be used for fracture assessment and

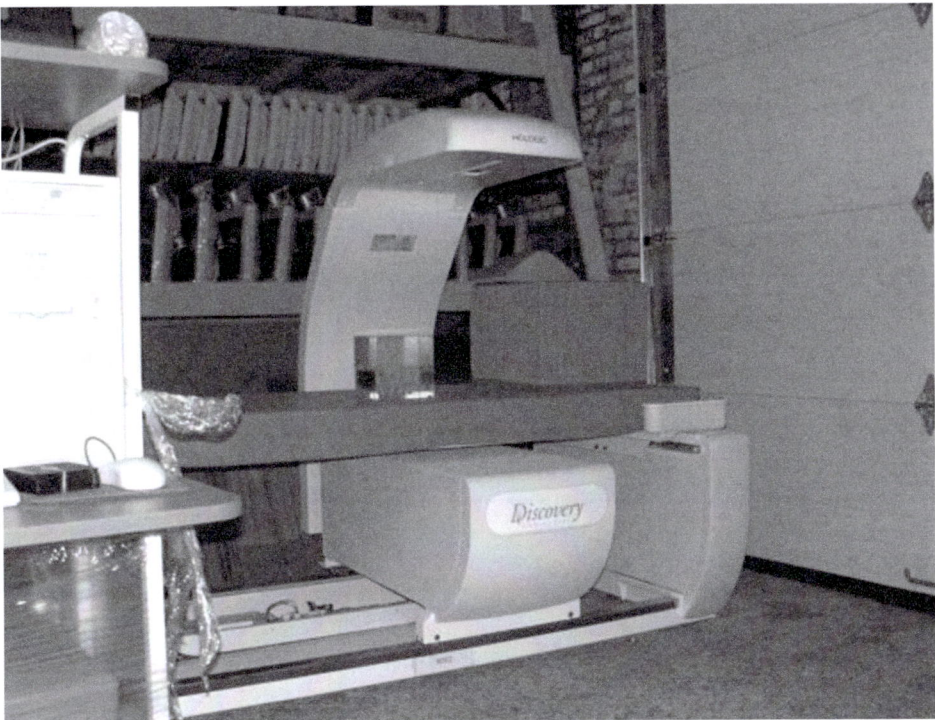

Figure 8.8 Dual-Energy Absorptiometry (DXA) is a means of measuring bone mineral density (BMD). Two X-ray beams with differing energy levels are aimed at the patient's bones. When soft tissue absorption is subtracted out, the BMD can be determined from the absorption of each beam by bone. The DXA scan is typically used to diagnose and follow osteoporosis. Image courtesy: Discovery Medical Inc.

thereby permit longitudinal studies of bone in space. By using this system, astronauts and ground-based flight surgeons will be able to accurately monitor rates of bone loss but, while having the ability to monitor bone loss will undoubtedly be helpful, the availability of a measuring system does nothing to reduce the loss of bone. To achieve this, countermeasures will need to be implemented.

Pharmacological countermeasures

The most common countermeasures to bone demineralization may be broadly classified into pharmacological intervention and non-pharmacological intervention. Pharmacological intervention includes the use of osteoporosis drugs such as *alendronate* (marketed under the brand name Fosamax^TM) and *calcitonin* (marketed under the brand name Miacalcin^TM). Both are approved by the Federal Drug

Administration (FDA) but, unfortunately for astronauts and flight surgeons, while the claims made for their efficacy suggest they have potential in bone-loss prevention, there are a number of drawbacks to their use. For example, alendronate, while very effective in promoting bone mass, must be taken for several years to gain the maximum benefit and the side effects of such long-term use are completely unknown. The other option, calcitonin, is a hormonal drug that results in bone mass gains of less than 2% a year – a figure far short of the required gains needed to offset losses during a multi-year mission. A more controversial drug is Slow Release Sodium Fluoride (SRSF), a formula that boosts the efficiency of bone-building osteoblasts but requires patients to have an annual blood fluoride check to ensure the drug stays below toxic levels in the body!

While calcitonin, alendronate, and SRSF may not be the drugs of choice for interplanetary astronauts, a more promising formulation is Osteoporex^{TM}, unique sea-algae calcium that is 90% absorbable by the body. Backed by more than a decade of research involving more than 300 treatment studies, the supplement has proved successful in promoting bone mass in 95% of the studies. An all-natural nutrient supplement that is four times as effective as synthetic pharmaceutical drugs, Osteoporex^{TM} may, in conjunction with other counter-measures described here, prove to play an important role in keeping an interplanetary astronaut's bones strong.

Now, you might be wondering why astronauts simply don't increase their calcium intake. After all, part of the reason bone mass is lost is because calcium is lost from the body. Well, it may seem a simple solution to the problem, but the remedy is a little more complex. Simply adding more calcium to the astronaut's diet wouldn't help and might even make the problem worse because excessive dietary calcium disrupts the delicate mineral balance needed by the body to repair and build bone. When the body's mineral content is over-weighted in favor of one particular mineral, the vital mineral balance is thrown off and it becomes more difficult to utilize *any* of the minerals properly. Because of this, researchers have directed their attention on proper calcium absorption and have discovered calcium balance can be maintained if it is used in small doses, in a highly absorbable form, and in proper balance with other absorption-promoting nutrients that enhance calcium metabolism. However, while this might help, it still won't be enough to help astronauts embarked upon multi-year missions. To further protect crewmembers, it will also be necessary to use non-pharmacological intervention strategies.

Non-pharmacological countermeasures

On Earth, non-pharmacological strategies such as exercise combined with adequate calcium, Vitamin D, and protein intake will maintain and even increase bone mass. A daily calcium intake of 0.1–1.0 g/day combined with a Vitamin D intake of 800 IU/day has been proven to reduce the risk of fracture. But, for interplanetary astronauts, more aggressive supplementation will be required. For example, on Earth, humans need about 200 IU/day of Vitamin D for bone growth, which most people achieve from exposure to sunlight. However, astronauts will be stuck inside a spacecraft and will need to maintain their Vitamin D levels by other means. Studies conducted on board nuclear submarines suggest that, in the absence of exposure to sunlight, Vitamin D intake should be boosted to between 500 and 600 IU/day. Another method by which Vitamin D may be maintained is by providing astronauts with a lighting system emitting the required amount of ultraviolet (UV) radiation to stimulate Vitamin D production.

In addition to taking drugs and increasing their calcium and Vitamin D intake, astronauts may benefit from eating fish oil. In 2010, NASA scientists found that by adding an omega-3 fatty acid called eicosapentaenoic acid (EPA) to regular bone cell cultures, the activation of factors that lead to bone breakdown was inhibited. The key factor that leads to bone loss (and muscle loss) is known as "nuclear factor kappa B" (a protein complex that controls the binding of DNA) or NFKB. Based on their findings, the scientists evaluated bone loss in astronauts and compared the results to reported fish intake during spaceflight. It was found that astronauts who ate more fish lost less bone mineral after four-to-six-month spaceflights. While it may be premature to conclude that the solution to the problem is simply a matter of diet, the scientists did find there was a link between the numbers of times astronauts ate fish in flight and the amount of bone they lost after flight.

Exercise

While drugs and nutrition will no doubt help astronauts maintain their bone strength, there is another countermeasure that has been used almost as long as there have been astronauts: exercise. You've probably seen astronauts running on treadmills on board the ISS and, if you're a fan of Stephen Colbert, you'll probably remember the fanfare accompanying the installation of the new treadmill (Figure 8.9) named after the talk-show host.

Officially called the Combined Operational Load-Bearing External Resistance Treadmill (COLBERT), the ISS's new $5 million treadmill got its name as a consolation prize for comedian, Stephen Colbert, who won an online NASA contest for the naming rights to a new space station module. In the competition, Colbert accumulated 230,539 votes, beating the runner-up, "Serenity", by 40,000 votes, but NASA ignored the votes (so much for fair competition!) and decided to name the module "Tranquility". Despite the boos from the audience when the decision was announced, Colbert was soon appeased when his guest, astronaut Sunita Williams, told him his name would instead grace the COLBERT treadmill! The COLBERT was duly unpacked on the orbiting station in September 2009 and it took astronauts about 20 hours to assemble from more than 100 pieces. After conducting a series of tests to make sure it was working properly, astronauts started running on the COLBERT and pronounced it a big improvement on the previous treadmill.

But, despite several hours spent exercising every day, astronauts still return to Earth with reduced bone density, although there are large differences between

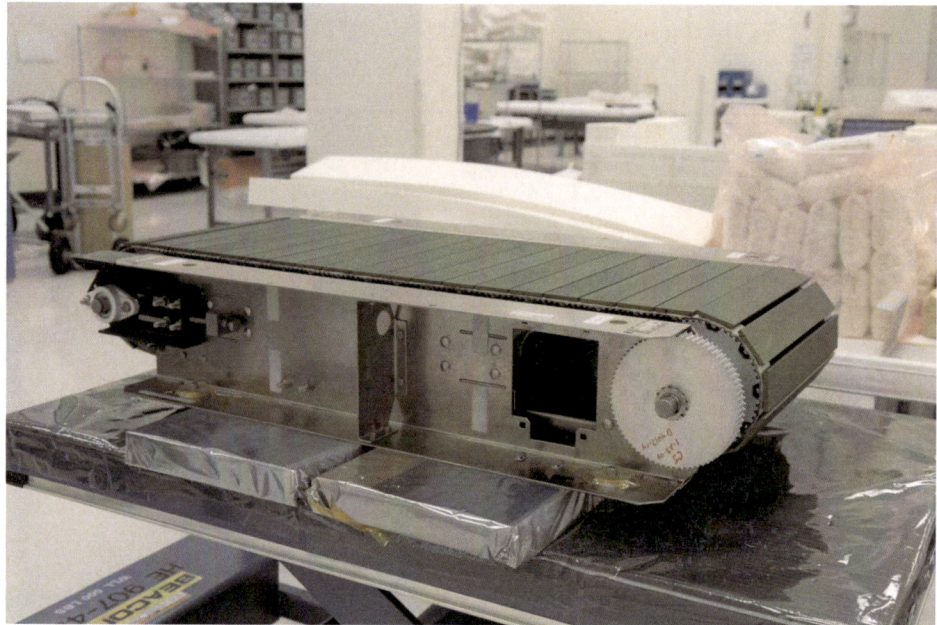

Figure 8.9 One small step for NASA, one giant running leap for Stephen Colbert. The Combined Operational Load Bearing External Resistance Treadmill (COLBERT). Image courtesy: NASA.

individuals – at one extreme, there is the case of NASA astronaut David Wolf, who spent 128 days on board Mir and lost up to 12% of his bone mass in certain areas, and at the other extreme is the case of cosmonaut Yuri Romanenko, who spent 326 days in space but didn't show any significant bone loss. In fact, Romanenko stood up unaided following the landing and ran 100 m the next day! But the cases of Wolf and Romanenko are the exception. Despite four decades of investigating ways to prevent bone loss, no astronaut has ever gone into space and not suffered bone loss. But bone isn't the only thing that wastes away in microgravity – astronauts also have to contend with muscle loss.

Muscle atrophy

The atrophy of muscles in space can not only affect the performance of astronauts during missions, but can also lead to severe muscle injuries upon return to Earth. The exact cellular and biochemical events that produce these losses of mass and strength are not well understood. It's why NASA and other space agencies such as ESA are conducting research on board the ISS to try to pinpoint the structural and metabolic changes that occur within individual muscle fiber cells and to better understand the effects of microgravity on the muscular system. One such experiment is the Muscle Atrophy Research and Exercise System (MARES) (Figure 8.10).

Figure 8.10 The Muscle Atrophy Research and Exercise System (MARES) is a general-purpose instrument intended for muscular and exercise research on the International Space Station. The device is capable of assessing the strength of isolated muscle groups around joints by controlling and measuring relationships between position/velocity and torque/force as a function of time. Image courtesy: NASA.

MARES is made up of an adjustable chair and human restraint system, a pantograph (an articulated arm supporting the chair, used to properly position the user), a direct-drive motor, experiment programming software, a linear adapter that translates motor rotation into linear movements, and a vibration isolation frame. The versatile device is capable of supporting measurements and exercise on seven different human joints, encompassing nine different angular movements, as well as two linear movements (arms and legs). It is considerably more advanced than current ground-based medical dynamometers (devices used to measure force or torque) and a vast improvement over existing ISS muscle research facilities.

Such a multi-purpose system enables scientists to study the detailed effects of microgravity on the human muscle-skeletal system and also provides a means to evaluate countermeasures designed to mitigate the negative effect on muscle, especially muscle atrophy. When in use, the crewmember taking part in an investigation sits on the MARES seat and operates the system using the MARES laptop following a set of computer-based instructions known as the MARES Experiment Procedures, which are made up of a sequence of prompts, MARES Profiles, displays of data, and steps to select data for storage and/or downlink.

Going blind in microgravity

While bone demineralization and muscle atrophy have been concerns as long as there have been astronauts, there is now a new microgravity-related problem spacefarers must contend with. A newly discovered eye condition has been observed in some astronauts who've spent months aboard the ISS and it has doctors worried that future explorers could go blind by the end of long missions.

While blindness is the worst-case scenario, the threat of blurred vision is enough for NASA to have asked scores of researchers to study the issue and put special eyeglasses on the ISS to help those affected see what they're doing. Given the potential for the vision problem to impact missions, NASA is treating the eye problem with a great deal of respect and considers the problem comparable to the risks already discussed in this chapter such as bone demineralization and muscle atrophy. According to one NASA survey, about 35% of crewmembers who have completed six-month shifts aboard the ISS reported a gradual blurring of eyesight. The disorder, similar to an Earth-bound condition called *papilledema*, is believed to be caused by increased spinal-fluid pressure on the head and eyes due to microgravity, although the exact cause is uncertain. Usually, the problem goes away once an astronaut returns to Earth, but a recent study by the National Academies noted that there had been "some lingering substantial effects on vision" and that astronauts were "not always able to re-qualify for subsequent flights". In one case, an astronaut never regained normal vision.

For decades, the space agencies had heard anecdotal evidence of vision problems, but scientists only began studying the issue in 2005 when an unnamed astronaut came forward. Much of the research to date has focused on seven unnamed astronauts who have shown symptoms, including one astronaut whose eyesight was so affected by his third month on the ISS that he could only "see the Earth clearly while looking through the lower portion of his reading glasses", according to an article published in the medical journal *Ophthalmology* [2]. In the months after returning to Earth, the astronaut noted a "gradual but incomplete improvement in vision", the report stated.

The seven cases were reviewed at a conference in February 2011 by dozens of scientists and doctors who concluded that the condition wasn't damaging enough to cause blindness in the near term but it was unknown whether it could be a long-term effect. The problem, according to many of the scientists, is that no one has been in space long enough to know just how bad the vision problem can get. It's a worrying prognosis, since, on Earth, papilledema can lead to blindness if left unchecked.

Obviously, you can't send astronauts to Mars if they're blind when they arrive, so research continues to see whether there is a countermeasure to solve the unusual eye problem. One common denominator may simply be age, since the astronauts participating in the study were all around the age of 50 and, in astronauts over the age of 40, like non-astronauts of the same age, the eye's lens may have lost some of its ability to change focus. After all, in the space program's early days, most astronauts were younger, military test pilots who had excellent vision, whereas today's astronauts tend to be in their 40s or older, which may be why there has been an increase in vision problems. It's a good theory, but you have to remember that, back in the early days, astronauts didn't spend six months in space, so the theory doesn't hold much water. Other researchers suspect that abnormal flow of spinal fluid around the eye's optic nerve or changes in blood flow in the tissues behind the retina may be involved. Ophthalmologists think these changes may be triggered by the fluids that shift towards the head when astronauts spend long periods of time in microgravity. As part of the research investigating the vision problem, astronauts receive comprehensive eye exams and vision testing such as pre- and post-flight MRI, optical tests that magnify cross-sectional views of parts of the eye, and special photography that records images of the retina and back of the eyeball.

Long-term exposure to microgravity represents a problem characterized by all sorts of idiosyncrasies. While drug intervention, nutritional manipulation, and exercise help, none of these strategies alone represents a solution to the problem. Perhaps this is because scientists are approaching the problem from the wrong angle. Rather than treating the effects of zero-G system by system, why not treat the root cause of zero gravity by simply restoring gravity?

References

[1] Mullane, M. Riding Rockets: The Outrageous Tales of a Space Shuttle Astronaut. Scribner (January 24, 2006).

[2]   Mader, T.H.; Gibson, C.R.; Pass, A.F.; Kramer, L.A.; Lee, A.G.; Fogarty, J.; Tarver, W.J.; Dervay, J.P.; Hamilton, D.R.; Sargsyan, A.; Phillips, J.L.; Tran, D.; Lipsky, W.; Choi, J.; Stern, C.; Kuyumjian, R.; Polk, J.D. Optic Disc Edema, Globe Flattening, Choroidal Folds, and Hyperopic Shifts Observed in Astronauts after Long-Duration Space Flight. *Ophthalmology*, **118** (10), 2058–2069 (2011).

9 Artificial Gravity

"After lunch, from 1300 to 1600, Bowman would make a slow and careful tour of the ship – or such part of it as was accessible. Discovery measured almost four hundred feet from end-to-end, but the little universe occupied by her crew lay entirely inside the forty-foot sphere of the pressure hull.

Here were all the life-support systems, and the control deck which was the operational heart of the ship. Below this was a small 'space-garage' fitted with three airlocks, through which powered capsules, just large enough to hold a man, could sail out into the void if the need arose for extravehicular activity.

The equatorial region of the pressure sphere – the slice, as it were, from Capricorn to Cancer – enclosed a slowly rotating drum, thirty-five feet in diameter. As it made one revolution every ten seconds, this carousel or centrifuge produced an artificial gravity equal to that of the Moon. This was enough to prevent the physical atrophy which would result from the complete absence of weight, and it also allowed the routine functions of living to be carried out under normal – or nearly normal – conditions.

The carousel therefore contained the kitchen, dining, washing, and toilet facilities. Only here was it safe to prepare and handle hot drinks – quite dangerous in weightless conditions, where one can be badly scalded by floating globules of boiling water. The problem of shaving was also solved; there would be no weightless bristles drifting around to endanger electrical equipment and produce a health hazard.

Around the rim of the carousel were five tiny cubicles, fitted out by each astronaut according to taste and containing his personal belongings. Only Bowman's and Poole's were now in use, while the future occupants of the other three cabins reposed in their electronic sarcophagi next door. The spin of the carousel could be stopped if necessary; when this happened, its angular momentum had to be stored in a flywheel, and switched back again when rotation was restarted. But normally it was left running at constant speed, for it was easy enough to enter the big, slowly turning drum by going hand-over-hand along a pole through the zero-gee region at its centre. Transferring to the moving section was as easy and automatic, after a little experience, as stepping onto a moving escalator."

From *2001: A Space Odyssey*, by Sir Arthur C. Clarke (1969)

Figure 9.1 A torus-shaped space station. Image courtesy: http://longavita.at.ua/space_station3.jpg.

Artificial gravity (Figure 9.1) is a common technology in science-fiction movies, even though the technology is rarely explained. On board the USS *Enterprise*, in the *Star Trek* universe, artificial gravity is achieved by using "gravity plating" embedded in the starship's deck. Computer games have also used artificial gravity as a setting. For example, the classic computer game *Halo: Combat Evolved* is set on an artificial ringworld that creates artificial gravity by computer-controlled rotational spin (inspired by Larry Niven's *Ringworld* that featured a habitat that created artificial gravity through rotation). Then there's the science-fiction classic *2001: A Space Odyssey*, which features a rotating centrifuge in the *Discovery* spacecraft, as described in the quote above.

When it came to creating artificial gravity, Sir Arthur C. Clarke got it absolutely right. Unlike so many other science-fiction writers who get around the nuisance of not having gravity by throwing a switch, the great Sir Arthur designed his space stations and spacecraft to produce artificial gravity for real. It's one of the reasons (and there are too many to mention here) that Kubrick's timeless classic, *2001: A Space Odyssey*, still stands head and shoulders above any film made since as the most dramatically realistic space movie ever conceived. And, as far as spacecraft go, the *Discovery* is unmatched not only in its aesthetic appeal, but also in its realistic functionality. But artificial gravity is no science-fiction gimmick. It's very real, which is why one of the most interesting aspects of the *Discovery* was that its artificial gravity generator could actually work, and here's how.

Wheels in the sky

The idea of spinning space stations and spaceships actually predates *2001: A Space Odyssey*. Back in 1945, Wernher von Braun, of Saturn V fame, prophesized the construction of space stations in orbit. His design, a toroidal station, which would be spun to provide artificial gravity, was popularized in *Collier's* magazine in 1952.

In the 1952 version, the station was 75 m in diameter and housed 80 crewmembers. The station's toroid was a smooth, donut-shape of inflatable sections made of reinforced rubber – a design that would utilize much new technology and would take at least 10 years to develop and build. The station would rotate to produce 1 G of artificial gravity at the crew levels. The outermost level would be dedicated to utilities and there would be two living and working levels. Small space taxis would be used to move from docking ports at the center of the rotating station to arriving space shuttles and to conduct assembly operations of Moon-bound spacecraft near the station. Sections of the station would be dedicated to communications, meteorology, terrestrial and military observations, and environmental control systems. The station would be positioned in a 1,730-km orbit (this orbit was later found to be within the then-unidentified Van Allen radiation belts and therefore unusable by a manned spacecraft). The orbit would be pseudo-Sun-synchronous, coasting along the terminator, perpetually in sunshine as Earth revolved below it (the space taxis would be used to adjust the orbit daily to keep it in this position throughout the year). The Sun-facing side of the station would be equipped with a solar concentrator trough, parabolic in cross-section, which would heat tubes containing mercury that would drive a turbine to generate electricity, then be cooled, condensed, and pumped back to the concentrator (a nuclear reactor power source was ruled out on grounds of weight).

The station would have a total volume of 18,400 cubic meters and require 24 metric tonnes of a nitrogen/oxygen air mixture for pressurization. Use of a helium/oxygen atmospheric mixture would reduce the total mass of atmosphere aboard the station to 16 metric tonnes and also eliminate the risk of bends in case of depressurization. The crew of 80 would require 110 kg of oxygen per day, but the use of blue green algae to convert the carbon dioxide exhaled by the crew to oxygen would nearly eliminate the oxygen resupply requirement.

Exhaled and urinated water would also be purified and reused. All these measures would reduce resupply requirements to 225 kg/day for the entire 80-person crew. Reserve supplies for 6–8 weeks would be replenished and kept aboard at all times. In case daily shuttle service was interrupted by an attack at the launch base on Christmas Island, as little as three launches every four months, each with a 33-metric-tonne payload, would be sufficient to resupply the station.

A rotating space station such as that devised by Wernher von Braun would produce the feeling of gravity on its inside hull. That's because the rotation drives any object inside the space station towards the hull, thereby giving the appearance of a gravitational pull directed outward. This "pull" is often referred to as a centrifugal force, but the pull is actually a manifestation of the objects inside the space station trying to travel in a straight line due to inertia. The space station's hull provides the centripetal force required for the objects to travel in a circle (if they continued in a straight line, they would leave the confines of the space station). Thus, the gravity felt by the objects is simply the reaction force of the object on the hull reacting to the centripetal force of the hull on the object, in accordance with Newton's third law.

Coriolis effects

From the perspective of astronauts inside the space station, artificial gravity by rotation behaves similarly to normal gravity but has some specific effects. First, there is centrifugal force, which, unlike real gravity which pulls towards a center, provides a rotational "gravity" that pushes *away* from the axis of rotation. Secondly, artificial gravity levels vary proportionately to the distance from the center of rotation so, with a small radius of rotation, the amount of gravity felt at the astronaut's head would be significantly different from that felt at their feet, which would make movement and changing body position awkward. Thirdly, there is the Coriolis effect, which exerts an apparent force that acts on objects that move. This force tends to curve the motion in the opposite sense to the habitat's spin, which causes problems for the body's balance system, since Coriolis effects act on the inner ear and can cause dizziness, nausea, and disorientation. Research has shown that slower rates of rotation reduce the Coriolis forces and scientists generally agree that a rotation rate of 2 revolutions per minute (rpm) won't produce any adverse Coriolis effects but, if the rotation rate is any higher, there are some people who won't acclimatize.

Engineering

Coriolis isn't the only issue faced by artificial gravity fans. Spinning up a space station requires energy and this requires a propulsion system, propellant, and a counterweight of some kind to spin in the opposite direction. Designing an artificial gravity habitat, whether it's a space station or a spacecraft, means extra strength is needed to keep the habitat from flying apart due to the rotation. Making matters worse is the fact that, if parts of the structure are intentionally not spinning, then friction and similar torques will cause the rates of spin to converge, which means more motors need to be fitted to compensate for losses due to friction.

The engineering challenges are part of the reason we have the clunky-looking

Figure 9.2 The International Space Station. Image courtesy: NASA.

International Space Station (ISS) orbiting Earth and not an elegant toroidal space station. As mentioned earlier, to reduce Coriolis forces to livable levels, a rate of spin of 2 rpm is needed and, to produce 1 g, the radius of rotation would have to be 224 m or greater, which would make for a very large and extremely expensive space station. The ISS (Figure 9.2) may look as if it was inspired by an Erector set but, at a cost of just over $100 billion, it was probably cheap in comparison to the effort that would have been required to design and build an Arthur C. Clark equivalent.

That's not so say engineers haven't tried to resolve the problem of designing an artificial gravity space station. One of the first steps was to scale down the design to see whether partial artificial gravity would work. To do this, it was suggested that to produce artificial gravity of 0.1 g would require a radius of only 22 m if the structure was rotated at 2 rpm. While 0.1 g is better than no gravity at all, it doesn't help astronauts combat occupational hazards such as bone demineralization and muscle atrophy, so scientists suggested the station have a radius of 10 m. This sounds like a small station and, if it spun at 2 rpm, it wouldn't produce any gravity at all, but the scientists got around that by spinning it at 10 rpm, which would actually produce Earth gravity (at the hips, but gravity would be 11% higher at the feet). The same design spun up to 14 rpm would produce 2 g. That sounds great until you remember that these rotation rates can only be tolerated for very short periods of time, but that's not to say the

Figure 9.3 The Agena Target Vehicle (ATV) was an unmanned spacecraft used by NASA during its Gemini program to develop and practice orbital space rendezvous and docking techniques and to perform large orbital changes, in preparation for the Apollo program lunar missions. By unreeling a 15-m nylon tether between the capsule and the Agena and flying in a dumbbell configuration with the Agena below, the astronauts were able to check the gravitational effect on the formation stability in uncontrolled mode. This technique is now known as gravity-gradient stabilization and uses a similar tether and a few thruster bursts to rotate the two craft around each other as an early test of artificial gravity. Image courtesy: NASA.

design is useless. After all, if brief exposure to high gravity can negate the health effects of weightlessness, then a small centrifuge could be used as an exercise area.

Given the potential benefits of artificial gravity, it's not surprising NASA has tested various configurations. One such attempt was conducted during the Gemini 11 mission, which produced artificial gravity by rotating the capsule around an Agena Target Vehicle that was attached by a 36-m tether (Figure 9.3). By firing the spacecraft's side thrusters to slowly rotate the combined craft like a slow-motion pair of bolas, the astronauts were able to generate a small amount of artificial gravity (about 0.00015 g).

The Stanford studies

While the resultant force was too small to be felt by either astronaut, objects were observed moving towards the floor of the capsule. After the Gemini 11, NASA didn't spend much time investigating artificial gravity until the subject came up during a NASA Summer Study conducted at Stanford University in the 1970s. During the Summer Study, which was convened for the purpose of speculating on designs for future space colonies, the Stanford Torus (Figure 9.4) was proposed.

Capable of housing between 10,000 and 140,000 permanent residents, the Stanford Torus would consist of a torus 1.8 km in diameter that rotated once per minute to provide between 0.9 and 1.0 g of artificial gravity on the inside of the outer ring. The original design had a tube with a width of 130 m that, with a tube circumference of 6.2 km, would give a living area for the inhabitants of 832,000 m^2. Sunlight would be provided to the interior by a system of mirrors. The ring would be connected to a hub via a number of spokes, which would serve as conduits for people traveling to and from the hub. Since the hub would be at the rotational axis of the station, it would experience the least artificial gravity and would be the easiest location for spacecraft to dock.

Not surprisingly, the Stanford Torus has inspired science-fiction writers. For example, the novels of the Gaea Trilogy by John Varley are set on an unusual organic satellite of Saturn that is shaped like a Stanford Torus and James P. Hogan wrote several novels that included a Stanford Torus, including *The Two Faces of Tomorrow*, *Endgame Enigma*, and *Voyage from Yesteryear*.

A habitat similar to the Stanford Torus is the Bernal sphere, first proposed in 1929 by John Desmond Bernal. Bernal's original proposal described a hollow spherical shell 1.6 km in diameter, with a target population of 20,000–30,000 people. The idea was refined during the studies at Stanford University by space visionary Gerard K. O'Neill, who proposed *Island One*, a modified Bernal sphere with a diameter of 500 m rotating at 1.9 rpm to produce a full Earth artificial gravity at the sphere's equator.

O'Neill's version would be capable of providing living and recreation space for a population of approximately 10,000 people, with a "Crystal Palace" habitat

Figure 9.4 The Stanford Torus is a proposed design for a space habitat capable of housing 10,000–140,000 permanent residents. It was proposed during the 1975 NASA Summer Study, conducted at Stanford University, with the purpose of speculating on designs for future space colonies. It consists of a torus 1.8 km in diameter and rotates once per minute to provide between 0.9 and 1.0 *g* of artificial gravity on the inside of the outer ring via centrifugal force. Image courtesy: NASA.

used for agriculture. Sunlight was to be provided to the interior of the sphere using external mirrors to direct it through large windows near the poles. In Figure 9.4, the residential area is located within the spherical portion. The structure at the two ends of the axial portion includes docking areas and the sites of zero-G manufacturing, while flat, paddle-like fixtures radiate away the waste heat of the habitat into space. Nearer to the sphere, the stacked rings are agricultural areas, helping to provide for the needs of the workforce. Here, agricultural crops, presumably genetically engineered to be radiation-resistant, are grown in the intense sunlight of space. The slightly curved plates, arranged in a circle surrounding the sphere, are the second stage in a series of mirrors that bring sunlight into the habitat at controlled hours. At a locus outside the figure, the mirrors of the first stage govern the "day–night" cycle. The second-stage mirrors pass the light at a desired angle to the ring-shaped mirrors capping the sphere. From these, sunlight is directed into the sphere's interior.

Centrifuge Accommodations Module

Incidentally, the report that NASA published in 1977 following the Stanford studies postulated that habitats of this type would be technically feasible towards the end of this century, possibly by the early 1990s. Pity they were wrong! After all, if astronauts had artificial gravity, they wouldn't have to worry about bone demineralization, muscle atrophy, or going blind. The problems encountered during long-duration space missions were part of the reason the ISS was to have had a module capable of simulating artificial gravity. Dubbed the ISS Centrifuge Accommodations Module (CAM) (Figure 9.5), the CAM is a cancelled element (along with the Habitation Module and Crew Return Vehicle) of the ISS. Although the module, which would have been docked next to the Harmony module, was planned to contain more than just a centrifuge, the 2.5-m centrifuge was considered the most important capability of the module. It would have provided controlled acceleration rates (artificial gravity) for experiments and the capability to expose biological specimens (providing they were less than 0.62 m tall) to artificial gravity levels of between 0.01 and 2 g. The centrifuge was also designed to provide a partial-g and hyper-g environment for specimens to investigate altered gravity effects and g-thresholds, and provide

Figure 9.5 The Centrifuge Accommodations Module. Image courtesy: JAXA.

short-duration and partial-g and hyper-g environments for specimens to investigate temporal effects of gravity exposure. Built by the Japanese Space Agency (JAXA), but owned by NASA, a CAM flight model was actually manufactured. It's now on display in an outdoor exhibit at the Tsukuba Space Center in Japan.

Artificial gravity initiatives

But interest in artificial gravity didn't die with the CAM. The next major initiative was the Mars Gravity Biosatellite, a project initiated as a competition between universities in 2001 by the Mars Society. Among the participating universities were the University of Washington and the Massachusetts Institute of Technology (MIT). The project's intent was to build a spacecraft to study the effects of Mars-level gravity (~ 0.38 g) on mice. The mission was to launch 15 mice into low-Earth orbit (LEO) for five weeks. The satellite was designed to spin at approximately 32 rpm to generate centrifugal force simulating gravity that astronauts would experience on the surface of Mars. At the end of its mission, the satellite would re-enter Earth's atmosphere and its cargo of mice would be retrieved. In 2007, a tentative launch date for the Mars Gravity Biosatellite had been set for 2010 or 2011, as the primary payload on a Falcon IE or a Minotaur IV launched from Cape Canaveral, Florida. In November 2009, MarsDrive took ownership of the project, and asked for support in the form of funding and research-and-design assistance. Sadly, that was as far as the project got.

Following the death of the Mars Gravity Biosatellite project, interest in artificial gravity waned. Then, in 2011, Mark Holderman and Edward Henderson of NASA's Johnson Space Center (JSC) proposed the Multi-Mission Space Exploration Vehicle (MMSEV) in a presentation to the Future in Space Operations (FISO) group. The MMSEV (Figure 9.6) was the result of some out-of-the-box thinking by the Technology Applications Assessment Team (TAAT), which had been examining key technologies that could advance space exploration.

The spacecraft, which also sports the designation of Nautilus-X (Non-Atmospheric Universal Transport Intended for Lengthy United States Explora-tion), isn't exactly easy on the eyes but it represents exactly the sort of divergent thinking that many people believe NASA should be engaged in. In short, the MMSEV/Nautilus-X is a proposal for a long-duration crewed space transport vehicle that includes a rotational artificial gravity space habitat intended to promote crew health for a crew of up to six on missions of up to two years. The partial-g torus-ring centrifuge (Figure 9.7) would utilize both standard metal-frame and inflatable spacecraft structures and would provide 0.11–0.69 g if built with the 12-m-diameter option.

The spacecraft depicted in Figure 9.6 is proposed to be relatively cheap by space system standards, as it is projected to only cost US$3.7 billion (by

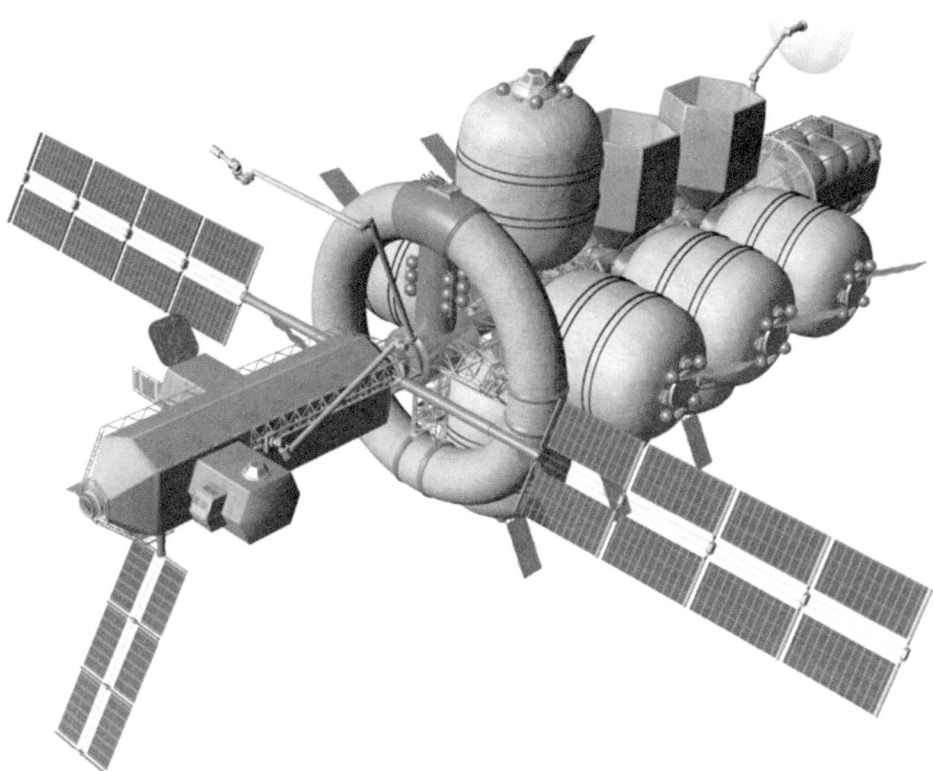

Figure 9.6 Mark Holderman's Multi-Mission Space Exploration Vehicle.
Image courtesy: Mark Holderman/NASA.

comparison, at the time of writing, NASA's Multi Purpose Crew Vehicle has soaked up more than $5 billion and it hasn't even flown). Going from aft to bow, we can see the propulsion system ringed by fuel tanks and radiators. Forward of the propulsion system are the expandable modules, containing logistical stores, and hangars containing the descent vehicle. Above the propulsion system, you can see the primary communications dish and forward of that are the EVA-pods and science probe craft. The environmental closed life-support system (ECLSS) is the vertical expandable module perched above the forward logistical store. Forward of this is the distinctive shape of the centrifuge and the photovoltaic (PV) arrays. Finally, at the sharp end of this visionary spacecraft is the command and observation deck, primary docking port, and the adaptable full-span remote manipulator system (RMS). Those with an engineering background may notice the lack of a thermal rejection capability. This will be resolved by the tried-and-tested classic basting roll maneuver, although radiators will probably still be required, along with some new variable conducting heat-pipe system. Engineers

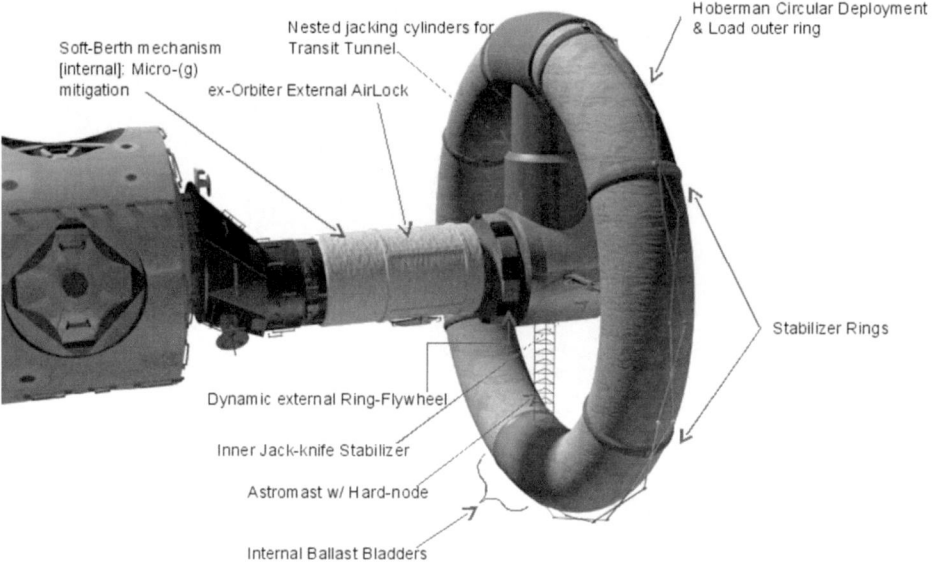

Soft-Berth mechanism [internal]: Micro-(g) mitigation

ex-Orbiter External AirLock

Nested jacking cylinders for Transit Tunnel

Hoberman Circular Deployment & Load outer ring

Stabilizer Rings

Dynamic external Ring-Flywheel

Inner Jack-knife Stabilizer

Astromast w/ Hard-node

Internal Ballast Bladders

Figure 9.7 The Multi-Mission Space Exploration Vehicle centrifuge. Image courtesy: NASA.

may also notice the absence of an exo-truss, necessary for managing and transmitting the load path from the propulsion system. That is because the issue of load path negotiation has yet to be resolved in this design. But let's get back to the artificial gravity. The centrifuge will be tested as part of a life science project at JSC. At JSC, scientists will test the partial-*g* effects on the human body and investigate how astronauts sleep at different *g* levels, how digestive function is affected, and how constant spinning affects mood and psychological function. Most likely, the centrifuge will be coupled to a sleep chamber, enabling astronauts to achieve maximum benefit from the fractional-*g*.

Once the centrifuge has been tested on Earth, it may be flown to the ISS (Figure 9.8) for more testing and evaluation. For example, engineers and mission planners will want to assess the effects of torque upon the ISS structure and ensure the drive mechanism works reliably. They will also want to ensure the means of deployment (remember, this is an *inflatable* module) and the stiffening load structures work as they should, while the crew will probably want to make sure that the partial-*g* toilet works more reliably than the one installed on board the ISS. Once the centrifuge has been tested, it would be ready for being attached to the MMSEV. As you can see in the diagram, the centrifuge would be constructed of several soft-wall inflatable sections and powered by start-up thrusters that would regulate the revolutions at a steady 10 rpm. It's envisaged the centrifuge demonstration project – if approved – will cost in the region of $84–134 million dollars and take about three years

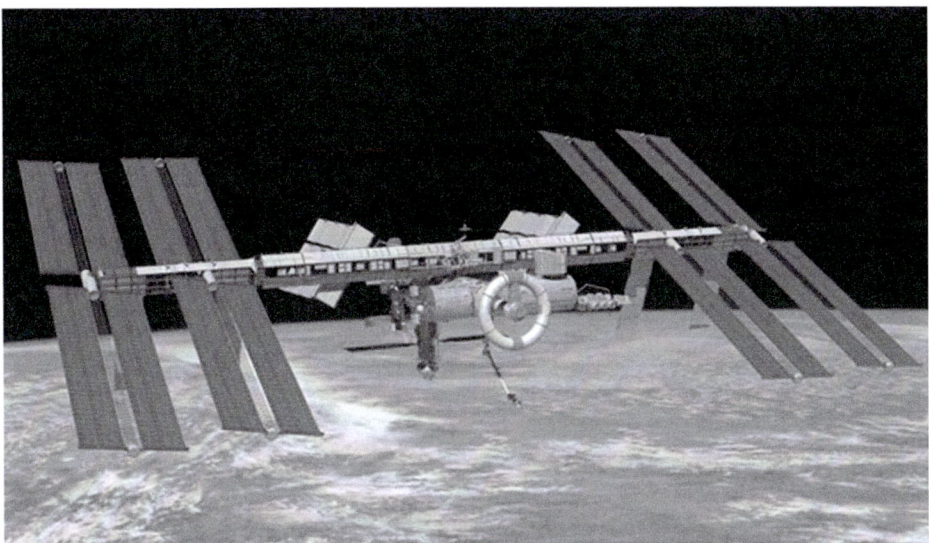

Figure 9.8 Inflatable centrifuge on the International Space Station. Image courtesy: Mark Holderman, NASA.

from design to implementation – a small price to pay to overcome one of the major challenges of interplanetary exploration and protecting the health of the crew.

If produced, this centrifuge will be the first in-space demonstration of a sufficient-scale centrifuge for artificial partial-g effects. At the moment though, the demonstrator is under study, with two diameters having been proposed (Table 9.1). If the demonstrator were to be flown, the centrifuge would be started using a kick motor similar to the Hughes 376 spin-stabilizers used on communication satellites. Eventually, the plan would be to reconfigure the centrifuge as a sleep module for the crew.

Table 9.1. Centrifuge.

Revolutions per minute	Partial-g 9.1-m diameter	Partial-g 12-m diameter
4	0.08	0.11
5	0.13	0.17
6	0.18	0.25
7	0.25	0.33
8	0.33	0.44
9	0.41	0.55
10	0.51	0.69

Artificial gravity research

While bean counters deliberate whether to fly the ISS centrifuge mission, scientists continue working on the problem of artificial gravity on Earth in the hope that one day their work will be flown. For example, Wyle Laboratories in Houston is working on the artificial gravity problem for NASA's Artificial Gravity program using a short-radius centrifuge (SRC). Known as the "Big AG", the SRC (Figure 9.9) is capable of spinning two riders simultaneously thanks to two arms that extend in opposite directions from a central pivot point. As the arms swing on the pivot, centrifugal forces create G-loads along the rider's body axis (head-to-feet) proportional to the rate of rotation. A device mounted on the footplate measures the G-forces at the feet, while other biomedical instrumentation mounted on the arms records heart rate and other physiological parameters.

For most of the tests, NASA spins the SRC at 17.3 rpm. At this speed, riders feel their feet pressing against the footplate a little more firmly than usual and notice some mild turbulence as air flows past the arm's windshield. When spinning in the dark, the SRC is so quiet that riders have very little sensation that they're moving at all, which is good news if astronauts end up sleeping in centrifuges. It may sound as if NASA is on its way to solving the problem of artificial gravity but, like any new technology, there are difficulties. One snag is the effect of rotation on astronauts as they spin. Spinning at 17.3 rpm on NASA's SRC is fine if you're spinning for only a few minutes but, as mentioned earlier in this chapter, after a

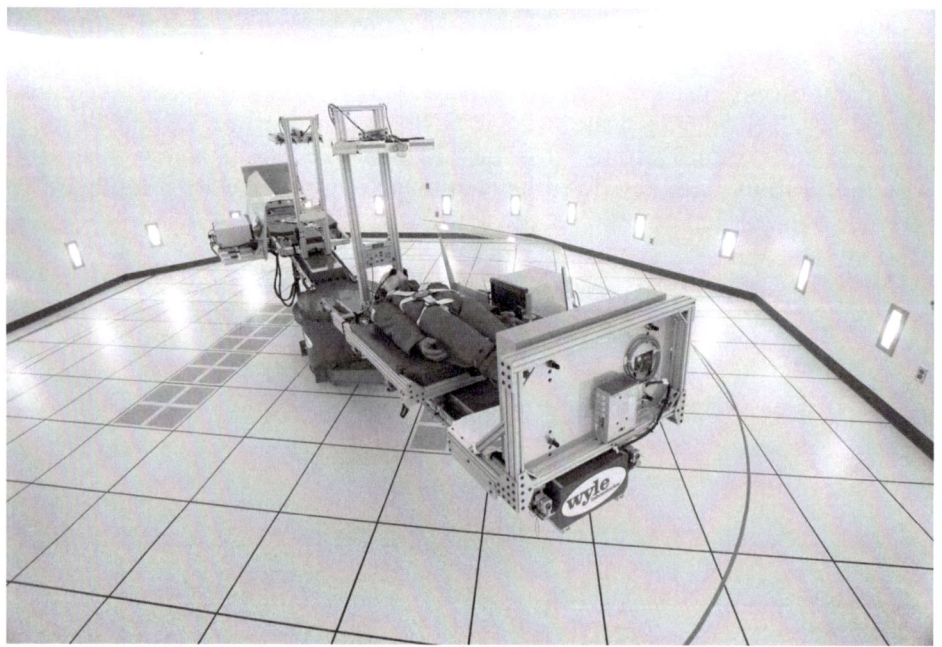

Figure 9.9 The "Big AG". Image courtesy: NASA.

while, some riders feel dizzy and disoriented. Of course, the solution is simply to reduce the spin rate, because it's known that a spin rate of 2 rpm or less produces no adverse effects, but the problem with such a low spin rate is that it doesn't produce sufficient gravity! It's a classic catch-22 situation. Even if NASA scientists solve the spin-rate issue, there's still the problem of angular movement to overcome. That's because high angular velocities produce high levels of Coriolis forces, angular moments (the amount of energy to spin) that would require a propulsion system of some kind to spin up (or spin down). Also, if parts of the spaceship are intentionally not spinning, friction and torque will cause the rate of spin to decrease as well as causing the otherwise stationary parts to spin. To compensate for these effects, flywheels and thrusters would be needed to keep certain sections of a spacecraft spinning or not.

It sounds complicated and it may take some time to resolve all the physiological and engineering challenges, so what other options are there? Well, there is another way to create artificial gravity and the idea is simple and within the scope of current technology: simply rotate (spin) the spacecraft and let centripetal force do the work for you.

Spinning spaceships

Before describing how spinning spaceships work, it's useful to know a few things about circular motion. To start with, remember that Newton's first law of motion states than any body in motion will remain in uniform motion (moving in a straight line at constant speed) unless a force acts on it. But, an object in circular motion must always be changing direction of motion, so there must be a force acting on it to make it move in a circle. This force is called centripetal force. Consider an astronaut standing in a large cylinder that is rotating. The outer edge of the cylinder, where the astronaut is standing, is moving at some speed and the cylinder walls must exert a force on the astronaut to keep him moving in a circle instead of flying off into space.

For the astronaut inside the cylinder, if they are standing still, all they feel is the wall of the cylinder pushing up on their feet. On Earth, gravity pulls down and, unless you feel the ground pushing up, then you feel weightless. This is the condition astronauts experience in orbit. Earth pulls down on them, but they are moving in orbit, always falling, but always missing Earth. So, they feel weightless. On solid ground, the ground pushes up on you with a force equal to your weight, which is the product of your mass times the acceleration due to gravity. But, inside the spacecraft, it turns out that if the astronaut is just standing there, they can't tell if it's level ground pushing them against gravity or it's the cylinder wall pushing on them to make them go in a circle. Pushing is pushing. If the astronaut wants to feel like they weigh the same as they did on Earth, the cylinder has to be spun fast enough for the centripetal force to equal the astronaut's weight. But that would mean spinning the cylinder rapidly, which would put a lot of engineering stress on it. So, perhaps you could get away with an apparent gravity of half of the astronaut's normal weight. Engineers calculate the acceleration due to gravity based on rotational speed by applying a little algebra, which is helpful if you want to know how fast to rotate the spacecraft to simulate gravity of a particular level. Of course, we don't know the minimum gravity needed for the human body to work properly over a long period of time – we'll have to wait for the NASA scientists to figure that out. What we do know is, the closer you want the fraction to be unity, the faster you have to rotate the spacecraft; and the bigger the radius, the slower the rotation. The obvious solution is to build a large spacecraft, but such a spacecraft will have a very large radius. Also, there is a lot of wasted empty space in the middle but the only part that has the desirable gravity is near the outer edge, so the obvious design would be a spacecraft shaped like a large torus. To provide a gravity environment similar to that on Earth would require a long-radius rotating vehicle up to 1 km wide. This would be prohibitively expensive, so let's look at alternate design options.

Perhaps the most novel solution to the artificial gravity problem is one suggested by Kent Joosten, who proposed designing spacecraft that spin. As you can see in Table 9.2, three spacecraft were proposed but the one judged the most feasible was the "Fire Baton" design (Figure 9.10), so we'll focus on that. The Fire Baton was designed to spin at a rate of 4 rpm. The spin rate was based on tests conducted at the Pensacola Slowly Rotating Room in the 1960s and 1970s. During these tests, it was determined that, at 4 rpm, some individuals would suffer no motion sickness, while others would adapt within a few days. Using a spacecraft with a rotational radius of 56 m, Joosten calculated that such a configuration would produce 1 g at 4 rpm.

As you can see, Joosten's spacecraft is an axis-spinner. The control jets located at the end of the arm mean they possess large moment arms, which in turn mean that, using a moderate level of thrust, the configuration would be spinning at the desired 4 rpm within two days. This spin rate would obviously generate high centripetal tension loads, so Joosten decided to use ultra-high modulus graphite for the spars due to the material's extreme stiffness. For the crew habitat, Joosten chose an inflatable structure attached to the end of one of the arms. Compared to spacecraft to date, the complexity of Joosten's spacecraft would be significantly reduced because there would be no need for microgravity systems. This would mean the waste disposal system, hygiene systems, and sinks wouldn't need vacuums to control free-floating debris, as is the case on board the ISS today. An additional feature of the 1-g crew habitat would be the inclusion of Earth-based comfort items such as chairs. No doubt, the habitat would include exercise devices, but these would be included more for crew relaxation than as a means to maintain bone density. While Joosten's design may offer an elegant solution to the problem of artificial gravity, the spinning spaceship is not without challenges, one of which is torque-gyroscopic precession. This is a phenomenon in which the axis of a spinning object "wobbles" when a torque is applied to it. The phenomenon is commonly seen in a spinning-toy top, but all rotating objects – including spaceships – can undergo precession. Needless to say, a wobbling spaceship would cause steering and navigation problems, and would

Table 9.2. Artificial gravity features of spinning spaceships.

Concept	Features	Potential advantages	Potential challenges
Fire Baton	• Hab counterweighted by reactor • Entire vehicle rotates • Vehicle pointing provides majority of thrust vector control (TVC)	• No rotating joints, power connections, fluid connections • Power conversion systems operate in *g*-field	• Vehicle angular momentum must be continuously vectored for TVC • Thermal radiators in *g*-field • Crew ingress/egress
Ox Cart	• Hab counterweighted by reactor • Thrusters despun, gimbaled for TVC	• Thrust vectoring decoupled from rotational angular momentum • Power conversion systems operate in *g*-field	• Megawatt-level power, prop transfer across rotating joints • Potential cyclical loading of rotating joints • Thermal radiators in *g*-field
Beanie Cap	• Split habitation volumes for counterweights • Reactor/power conversion systems, thrusters in zero-G • Thrusters gimbaled for TVC	• Thrust vectoring decoupled from rotational angular momentum • Thermal radiators in zero-G	• Inefficiencies in duplicating habitation systems, crew transfer between them • Potential cyclical loading of rotating joints • Power conversion systems operate in *g*-field

probably result in the crew feeling a little queasy. There are various ways to counter the problem, such as differential thrusting and cyclical torquing, which may reduce the precessional rate, but these solutions are theoretical. A second problem is the issue of center of gravity (CG) control, which may be caused by the CG offsets in the habitation and power modules. Any such offsets would cause stability problems, although they could be resolved using active ballasting.

If engineers *can* make artificial gravity work without making the crew sick or dizzy, it's possible that, one day, interplanetary astronauts will travel to the outer planets on spinning spaceships similar to that designed by Joosten. If the

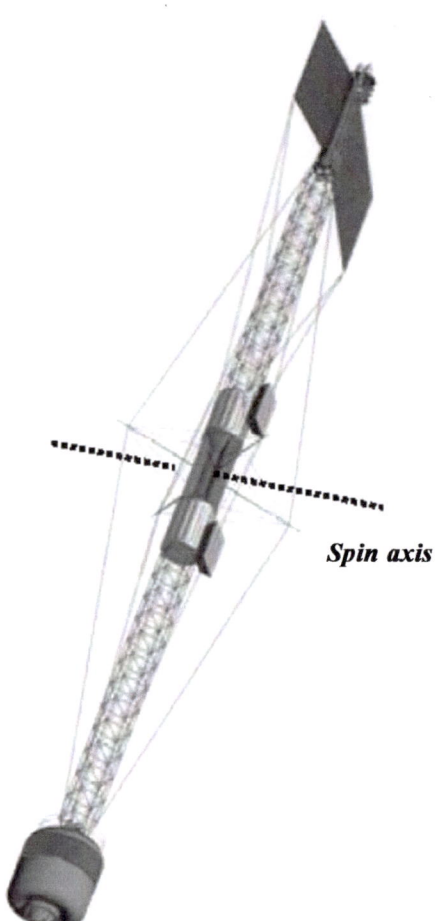

Spin axis

Figure 9.10 The "Fire Baton" spinning spaceship designed by Kent Joosten. Image courtesy: Kent Joosten/NASA.

engineers *can't* make it work, then perhaps the answer will be a combination of medical countermeasures and artificial gravity – a solution that would represent a combination of clever medicine and clever spacecraft. But, what if *none* of these methods works? What then? Well, perhaps those science-fiction ideas aren't so far-fetched after all.

Gravity generators

The artificial gravity generator is one science-fiction device particularly disliked by physicists. Used as a plot device as early as 1930 by Olaf Stapleton, artificially produced gravity fields make spaceflight a lot easier and more bearable for everyone. But it's impossible, right? Perhaps not. Recent work conducted by researchers supported by the European Space Agency (ESA) has measured the gravitational equivalent of a magnetic field for the first time in a laboratory. Martin Tajmar, Clovis de Matos, and colleagues have successfully produced and measured a very weak gravitomagnetic field by applying angular accelerations to a superconductor. The field produced is directly proportional to the applied acceleration with a correlation factor higher than 0.96. The gravitational field emitted from the superconductor followed the laws of field propagation and induction similar to those of electromagnetism as formulated in linearized general relativity. The study was the first time non-Newtonian gravitational and gravitomagnetic fields of measurable magnitude had been observed in a laboratory environment. Tajmar's team ran more than 250 experiments over three years and discussed the validity of the results for eight months before making the announcement. They are now waiting for other physicists to conduct their own versions of the experiment to verify the findings and rule out a facility-induced effect.

It's an intriguing development, especially if it can be duplicated by other researchers. If the results can be verified, then it puts the technology of science fiction on the horizon and would make long-duration spaceflight a much more comfortable and safe proposition; no need to spend hours every day running on a treadmill – just switch on the force field and let it do the work.

Appendix I
Human Centrifuges

For the last four years, I've been in charge of Canada's only human centrifuge, located at Defence Research and Development Canada (DRDC) in Downsview, Toronto. As an Acceleration Training Officer (ATO), I've trained and spun hundreds of pilots to help them test their reactions and tolerance to the dreaded "G". People often ask how one becomes an ATO. After all, since the number of human centrifuges in the world is less than a couple of dozen, my job is a rather unique one, which requires a unique set of skills. Basically, to be trained as an ATO requires a solid grounding in aerospace physiology and a comprehensive understanding of safety, G-training, personnel and roles, control room operations, recovery operations, pre-flight and post-flight debrief procedures, emergencies, and emergency pilot egress procedures. It also helps if you've ridden the centrifuge and if you understand how G feels in the back seat of a jet

Figure A1 AMST centrifuge; this one is located in China. Image courtesy: AMST.

Figure A2 AMST centrifuge; this one is located in Poland. Image courtesy: AMST.

so you can commiserate with the pilots when they step out of the centrifuge gondola and tell you that it's nothing like the Gs pulled in their F-18!

Since the human centrifuge is such an unusual item of training equipment, I thought readers might be interested to see some of the centrifuges that operate around the world and the companies that manufacture them.

The list of companies that manufacture man-rated centrifuges is a short one and includes Austria Metall SystemTechnik (AMST) in Austria and Environmental Tectonics Corporation (ETC) NASA in the US, although the space agency doesn't manufacture commercial "fuges".

You can see two of AMST's centrifuges in Figures A1 and A2. The one depicted in Figure A1 is now spinning in China and the one in Figure A2 is used by the Polish Air Force. Each costs in the region of $25 million. One of AMST's competitors in the centrifuge-building business is ETC, which is the company that built the centrifuge I operate, and, more recently, they built the new centrifuge at Wright Patterson Air Force Base.

Perhaps the most advanced centrifuge in the world is depicted in Figures A3 and A4. Thanks to several advanced technological upgrades that have resulted in an increase in simulator fidelity, the Phoenix is the most technologically advanced high-G flight trainer available. Capable of generating the variable G onset/offset rates and G-forces of an actual jet fighter to give pilots the most realistic training, the Phoenix gives pilots an experience that is almost the same as actually flying the aircraft. The Phoenix also allows pilots to learn and refresh

Figure A3 The Phoenix. Image courtesy: ETC/NASTAR.

Figure A4 The Phoenix (side view). Image courtesy: ETC/NASTAR.

Figure A5 The Gyrolab. Image courtesy: ETC/NASTAR.

Figure A6 The Gyrolab (side view). Image courtesy: ETC/NASTAR.

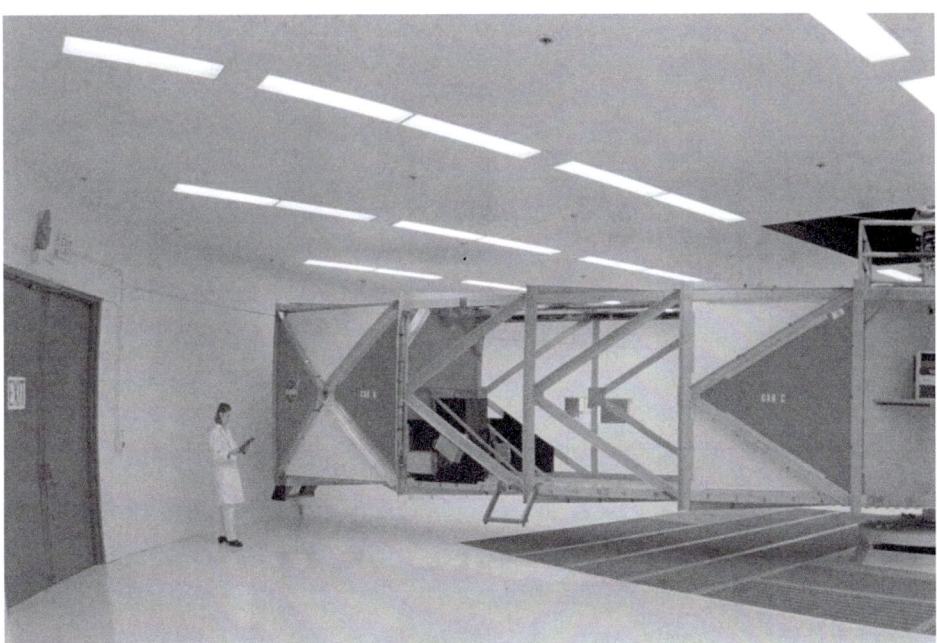

Figure A7 NASA's 20-G centrifuge. Image courtesy: NASA

Figure A8 One of the cabins attached to NASA's centrifuge. Image courtesy: NASA.

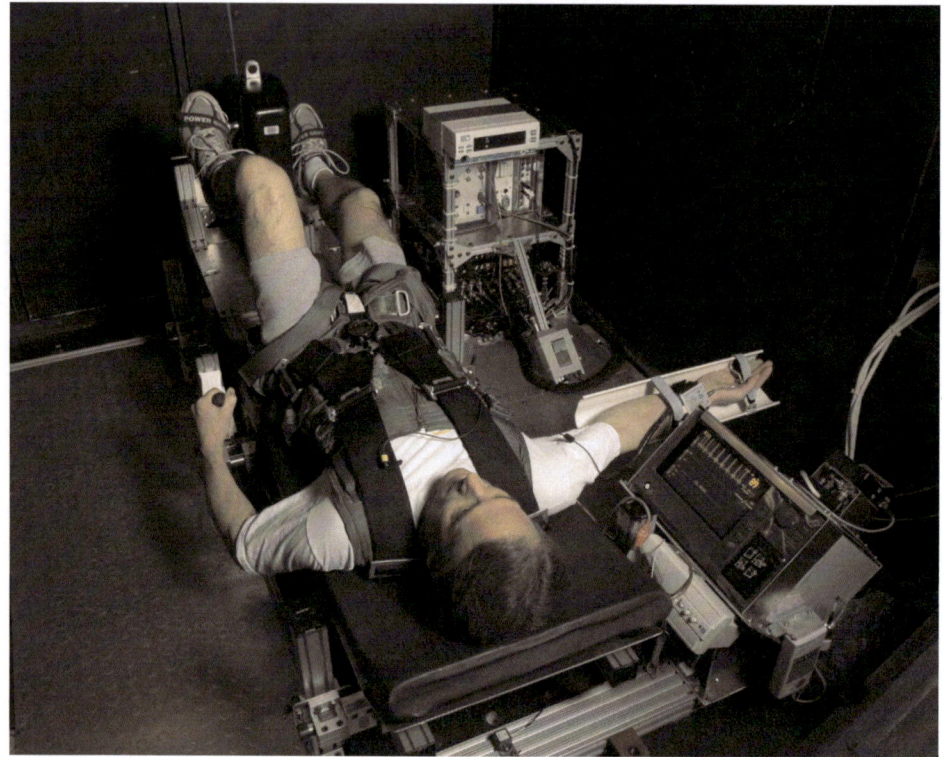

Figure A9 Russell Westbrook, Experiment Subject, rides NASA's 20-G centrifuge. Image courtesy: NASA.

their skills in an environment that includes the same stresses that they might encounter during actual flight in complete safety.

Another ETC product is the Gyrolab GL-4000 (Figures A5 and A6), a high-fidelity, single-seat interactive motion platform that provides users with 360° of continuous and simultaneous motion in four axes of motion (planetary, pitch, roll, and yaw). Up to 6 G of motion stimuli are generated in the planetary axis. The Gyrolab GL-4000 is used for flight training or research applications and provides a realistic flight-training environment with a full, high-fidelity cockpit of the client's choosing. An interactive flight profile editor allows the operator to pre-program flight scenarios for Spatial Disorientation (SD) and Situational Awareness (SA), which makes the unit perfect for flight training.

NASA scientists use a 20-G centrifuge (Figure A7) to evaluate the effects of hypergravity on humans. The main goal of these experiments is to reduce the adverse effects that space travel can have on astronauts' physical heath. The 17.5-m-diameter centrifuge has three cabins (Figure A8), one for humans – limited to 12.5 G – and two for objects and flying hardware.

One cab, mounted at one end of the rotating arm, contains a modified jet fighter seat in which a human sits during tests, as you can see in Figure A9.

Appendix II
Air Force Pamphlet 11-419 – G-Awareness for Aircrew

BY ORDER OF THE
SECRETARY OF THE AIR FORCE

AIR FORCE PAMPHLET 11-419

1 DECEMBER 1999
Certified Current, 29 JANUARY 2010

Flying Operations

G-AWARENESS FOR AIRCREW

NOTICE: This publication is available digitally on the AFDPO WWW site at: http://afpubs.hq.af.mil.

OPR: AFMOA/SGOA (Col Albert A. Hartzell)

Supersedes AFPAM 11-404, 19 August 1994

Certified by: AF/SG/SGM2
(Col Michael D. Parkinson)
Pages: 15
Distribution: F

This pamphlet implements AFPD 11-4, *Aviation Service*, and AFI 11-403, *Aerospace Physiological Training Program.* It provides all high-G aircrew with a source of reference for information and techniques, and covers basic physiology of high-G flight. It explains how to prevent G-induced loss of consciousness (G-LOC) and includes the effects of G-forces on the body, the factors that increase and decrease G-tolerance, and countermeasures to avoid G-LOC. This pamphlet applies to all high-G aircrew and is authorized for issue to each aircrew member. As a minimum, each unit will ensure the local Publications Distribution Office orders enough copies to make the pamphlet readily available to each aircrew member.

Send comments and suggested improvements through the parent MAJCOM Aerospace Medical Office, to AFMOA/SGOA 110 Luke Avenue, Suite 405, Bolling AFB DC 20332-7050.

SUMMARY OF REVISIONS

This revision renumbers the AFPAM, updates information on Class A mishap rates (paragraph 1), adds information about high intensity circuit training (paragraph A2.3.6.), aerobic conditioning (paragraph A2.5.) and deletes illustrations (paragraph A2.6.) Changed or revised material is indicated by a bar (|).

Section A—PHYSIOLOGY

1. Importance of G-LOC Avoidance.

1.1. G-LOC continues to be a very real threat in fighter aviation. Class A mishap information shows that 26 aircraft were lost to G-LOC between 1982 and 1998. Reported Class C mishap information shows 402 G-LOC between 1985 and 1998. Research in both physiology and hardware improvements continue to be pursued. Until our knowledge and technology provide the means to greatly reduce the impact and threat of high-G forces, survival is the responsibility of each aircrew, on each and every flight. An understanding of the basic physiology of G forces as they affect the heart, brain, and eyes is necessary to facilitate later discussions on factors that influence G tolerance.

2. Introduction to G-Forces.

2.1. An increase in gravitational (or "G") forces results in reduced blood pressure and blood flow to the brain and eyes. Without blood flow, these organs can only function for about 4 to 7 seconds on oxygen reserve (less if at greater than one G), after which the sequential effect is grayout, tunnel vision, blackout and loss of consciousness. With a rapid reduction in blood pressure (such as that seen with rapid G onset or with relaxation of straining at high G), these sequential effects may go unnoticed, and loss of consciousness can occur without visual warning.

2.2. With a fall in blood pressure, the cardiovascular system responds with a reflex increase in blood pressure and heart rate. The full reflex response takes 10 to 15 seconds to complete.

2.3. G stress greater than resting G tolerance (with or without the cardiovascular reflex response) requires an anti-g straining maneuver (AGSM) to maintain consciousness. Rapid G onset or relaxation of the straining maneuver at high G levels can lead to G-LOC without any visual symptoms or warning signs. After a G-LOC, the brain requires an average of 24 seconds (range 8 to 80 seconds) before awareness of the environment returns, and even greater time may be required before mental skills are back to normal.

3. Heart, Brain, and Eyes.

3.1. The heart is a pump, supplying oxygenated blood throughout the body. In a discussion of G stress and G tolerance, the brain and eyes are the most important organs to consider. Blood pressure, flow rate and vertical distance from the heart to the head all influence the heart's ability to supply adequate blood flow to the brain during G stress. Consciousness depends on blood being available to the brain while vision depends on blood supply to the retina of the eye. The eye requires a slightly higher blood pressure since it has a static inflation pressure (intraocular tension) which must be overcome. Both the brain and the eye have a 5 to 7 second reserve of oxygen, which will maintain function following sudden loss of blood flow.

4. Effect of G-Stress.

4.1. The most significant effect of positive G-forces on the brain and the eyes is the reduction of blood pressure and blood flow. The eye reacts first to the reduced blood pressure. As G-forces increase and blood pressure drops, aircrew experience grayout (loss of color and clarity), tunneling of vision, and then visual blackout. Visual blackout can occur without loss of consciousness, and this differentiation should not be confused. These visual symptoms reverse immediately upon restoration of blood flow and oxygenation to the eye.

4.2. When the brain loses its blood supply and exceeds its oxygen reserve, it abruptly fails. Once it fails, it stays "turned off" for a variable length of time, even after blood flow is restored.

4.3. Consciousness can be maintained when the G-onset rate is slow enough that visual symptoms are recognized and the AGSM is intensified accordingly. When G-onset rates are high and the peak G level sustained is high, as is often the case in modern fighters, G-LOC can occur without any visual warning signs when the oxygen reserve is depleted. Aircrew exposed to lower onset rates and peak G levels can also G-LOC if they ignore the visual symptoms, but should never use visual symptoms to test current G-tolerance. Mishap statistics and surveys have shown that no aircrew is immune to G-LOC susceptibility; it can happen to anyone flying any aircraft at any time when under G stress.

AFPAM11-419   1 DECEMBER 1999   3

5. Post G-LOC.

5.1. Initial Recovery. Once G-LOC has occurred, the brain enters a state called "absolute incapacitation" which lasts an average of 12 to 16 seconds (range 5 to 30 seconds). This will occur even if aircraft G has been unloaded. During this time, the individual is in a dream-like state, unaware of his or her environment and unable to respond to any outside stimuli. The brain's blood supply returns during this period, if G is reduced. After this absolute incapacitation period, the brain starts to "wake up" and enters a "relative incapacitation" period which lasts on average another 12 seconds (range 3 to 50 seconds). The combined absolute and relative incapacitation times are referred to as the total incapacitation time; this total period averages about 24 seconds (range 8 to 80 seconds). At the end of this total incapacitation period, the individual is able to recognize where he or she is and respond to the environment. Aircraft recovery may be possible in the "relative incapacitation" period, as the aircrew member may respond to directive calls such as "pull-up, pull-up".

5.2. Full Recovery. There is a third phase to recovery from G-LOC: the return of cognitive processing skills. This may require several minutes before a full return to function. During this time, BFM skills and situational awareness may be severely impaired. In addition, a sudden high-G attempt at aircraft recovery soon after a G-LOC may induce a subsequent G-LOC episode. The implications of G-LOC and its potential immediate after-effects in a combat scenario are obvious. You may be conscious after a G-LOC, but you won't be a full-up combat aircrew member.

5.3. G-LOC Recognition. G-LOC may go unnoticed due to a partial amnesia that can occur as a result of impaired oxygen flow to the brain. You should suspect that G-LOC may have occurred if you cannot explain a sudden loss of altitude or have difficulty recognizing your aircraft attitude. Tingling around the mouth or in the extremities or a sense of dreaming are just two of the various sensations you may experience upon recovery from G-LOC. It is possible that review of your head-up display (HUD) film during debriefing may be your first clue that you experienced an inflight episode of G-LOC. NEVER USE VISUAL SYMPTOMS TO TEST YOUR CURRENT G TOLERANCE. If you experience grayout under G, you have only a 10 percent oxygen reserve remaining and are dangerously close to going to sleep. Visual symptoms should be considered a last resort warning that your are near G-LOC.

Section B—G-TOLERANCE

6. G-Tolerance Factors.

6.1. G-tolerance is the ability or capacity to maintain vision, consciousness, and effective performance when under G-stress. To do this, blood pressure and flow must be maintained to the brain and eyes. Factors which affect an aircrew's G tolerance include: 1) the effectiveness of the anti-G straining maneuver; 2) anticipation of the G-onset, 3) aircrew physical, physiological and psychological characteristics; 4) G-protection equipment and 5) G-discipline and awareness. The straining maneuver is the critical component of high G-tolerance, but G-awareness and "G-discipline" for high onset rates and sustained G are the critical elements of G-LOC avoidance. For quick reference, these factors and their corresponding explanatory paragraphs are listed below.

7. Anti-G Straining Maneuver.
The AGSM is the best G-defense measure available to aircrew members. Equipment measures (anti-G suit, COMBAT EDGE, reclined seat, etc.) were never meant to replace the AGSM, only aid it. Aircrews must properly perform a timely AGSM in order to gain maximum ben-

efit. It is very important that you perform the same, correct AGSM each time you anticipate and or apply G, regardless of the amount of G. This will ensure positive feedback, and will imprint the proper AGSM to make it an automatic response.

7.1. Tolerance. Average aircrew relaxed G tolerance in the F-16 seat is about 5.2 G (about .5 to .75G less in aircraft without a reclined seat); the G suit can add another 1G, and a good AGSM can add another 3.5G or more of tolerance. When these are totaled, one can see that 9G is a big challenge for most aircrew; there is little or no safety margin.

7.2. Preparation. The straining maneuver should precede rapid G onset. It is extremely difficult to "catch up" to a G load if you get behind from the start. In addition, proficiency in performing the maneuver may decrease (decondition) when you have not flown recently. Physical conditioning, fitness, and mental preparedness, proficiency and currency are all keys to an effective AGSM.

7.3. Components. The AGSM consists of two components: maximal contraction of lower body muscles and forced exhalation against a closed glottis. Performing one without the other may significantly reduce the effectiveness of the strain.

7.4. Checklist.

7.4.1. Anticipate the G.

7.4.2. Simultaneously tense all lower body muscle groups legs, butt and abdominal muscles (maintain this strain)

7.4.3. Deep breath, with forced exhalation against closed airway.

7.4.4. Quick breath every 3.0 seconds, once at peak G.

7.4.5. Minimize communications.

7.4.6. Don't relax until G is unloaded.

7.5. Details.

7.5.1. Muscle Tensing. Muscle tensing increases usable blood volume and return of blood to the heart.

7.5.1.1. Begin with lower body muscle groups, legs, butt and abdominals; this reduces the space in which the blood may pool.

7.5.1.2. The tighter (more tense) the muscles the greater the reduction in the tendency for blood to pool.

7.5.1.3. The legs are most important.

7.5.1.4. Tensing of all skeletal muscles, especially abdominals, helps to raise blood pressure and "push" blood back to the heart.

7.5.2. Upper Body Muscular Strain.

7.5.2.1. The upper body muscle strain may be varied depending on the G-load, but the tenseness must continue until unloaded.

7.5.2.2. The muscle strain must be maintained continuously, even when breathing.

7.5.2.3. If the muscles are relaxed while still under G, the blood will immediately rush into the extremities (opened spaces). This makes it almost impossible to catch up if at moderate G (4 to 5) or higher and may even result in almost instantaneous G-LOC.

7.5.3. Increased Chest Pressure. Increased chest pressure increases heart output pressure. Works as a "boost pump" for the heart. The greater the pressure generated in the chest, the more the heart and blood vessels leading from it are squeezed, and the greater the resultant blood pressure. This keeps brain blood pressure in the functioning range.

7.5.4. Breathing Cycle.

7.5.4.1. Take a deep breath prior to the onset of G, close the glottis (throat), and bear down with the chest muscles as in trying to exhale, but keep the throat closed. This process will generate the pressure.

7.5.4.2. Every 2.5 to 3.0 seconds, take a breath. Exhale a small amount of the air. Immediately pull the air back in and regenerate the chest pressure. The exhalation and inhalation process should ideally take about 0.5 seconds.

7.5.4.3. Minimize communications during G.

7.5.5. Common Errors.

7.5.5.1. Having proper knowledge of the technique, *but* failing to turn that knowledge into a skill, which is integrated with other flying skills.

7.5.5.2. Developing good chest pressure, but failing to tense the lower body musculature. This causes the blood to pool in the extremities and the overpressure in the chest impedes the return of blood to the chest. The result of this counterproductive strain can be G-LOC or severe visual loss regardless of the G load.

7.5.5.3. Failing to anticipate the G. Performance of the AGSM should begin before G is loaded on the aircraft. Failure to do so will result in the aircrew member either trying to catch up on the AGSM (a very dangerous practice) or having to unload in order to buy time to catch up.

7.5.5.4. Failing to maintain chest pressure (loss of air). Occurs while talking or whenever the strain is audible. As air is lost from the chest the amount of pressure generated falls. This directly reduces blood pressure to the brain. If air loss is heavy, as might occur with speech, the subsequent loss of blood pressure in the brain may result in G-LOC without visual loss. Other causes of air loss are "groaning" (letting the air escape slowly) and trying to hold the chest pressure by sealing the lips rather than with the throat.

7.5.5.5. Holding the breath less than the required 2.5 to 3.0 seconds. Results in lower average blood pressure in the brain, and fatigue is accelerated.

7.5.5.6. Holding the breath longer than 3.0 seconds. The increased chest pressure impedes return of blood to the chest where it is available to the heart. If blood return to the chest is blocked for 4 to 5 seconds the heart may run out of blood to pump. This can be very dangerous when wearing COMBAT EDGE.

7.5.5.7. Taking too long to complete the exhalation and inhalation cycle. Total time for this cycle should be 0.5 seconds.

7.5.5.8. Performing a strain with the intensity necessary to stay awake at 9G when the G-load is only 5G, will result in early fatigue and increase potential for G-LOC in subsequent engagements. The intensity of the AGSM should be graded in relation to the level of G. It is always safe to overestimate the intensity of the strain, and always unsafe to underestimate the intensity required.

7.5.6. Training and Proficiency. While the aircraft and personal G-protection equipment are passive, the AGSM is active. It requires anticipation of the maneuver and is a practiced skill which must be integrated with numerous other cockpit tasks. The efficiency and intensity of the AGSM depends on multiple factors, including strength, endurance, training, motivation, and proficiency. The AGSM is like other athletic skills in that it is susceptible to deconditioning. Scientific studies indicate there is some decrease in the endurance of the AGSM when the aircrew member is not current and proficient. Tolerance to peak G for short intervals may not be affected, *but* ENDURANCE TO MULTIPLE OR SUSTAINED HIGH G ENGAGEMENT MAY BE DEGRADED AFTER A LAYOFF FROM HIGH G FLYING; ESPECIALLY IF ILLNESS WAS THE CAUSE!

7.5.6.1. The best time to practice the correct AGSM is during the G-awareness maneuver. This is a part of the sortie during which you can devote almost all of your conscious attention to G and to your AGSM technique. Once the engagements start, your attention will be directed toward the mission and practice of the AGSM will probably be a distraction. You should use hot mike during the G-awareness turn, so you can review your AGSM during the debrief. *Do not waste this opportunity!*

7.5.6.2. Centrifuge training has proven to be an invaluable tool in teaching aircrew a proper AGSM without worry about the consequences of a G-LOC in flight. The Combat Air Forces (CAF) require that all high-performance aircraft aircrew members go through centrifuge training. Training requirements and policy are described in AFI 11-404, *Centrifuge Training for High-G Aircrew.*

7.5.7. VTR Review.

7.5.7.1. There are several requirements that should be met in order to provide the best appraisal of an individual's AGSM through VTR review. The Flight Supervision, flight surgeon, or aerospace physiologist should perform the following during the VTR review:

7.5.7.2. The aircrew should be on "hot mike", and the HUD should be easily read.

7.5.7.3. The aircrew should be performing other tasks while under G. Ideal situations: BFM, ACT, etc.. Several engagements on a single sortie should be appraised.

7.5.7.4. The adequacy and correctness of tensing the extremities and other musculature cannot be judged. The automaticity of the strain cannot be adequately judged during "canned" maneuvers such as G-Awareness turns.

7.5.7.5. Listen for a preparatory inhalation just before the G is loaded. If it is not there, the aircrew may already be behind the G.

7.5.7.6. Listen for exhalation sounds or talking during the G-onset. This signifies loss of air from the chest and reduced efficiency of the strain and G-tolerance. Additionally, the aircrew is likely to be behind the G and will have trouble catching up. This may cause the aircrew to either unload some of the G or sharply increase the intensity of the strain (usually audible). The latter may be a result of recognition of vision loss. Ideally, the first breath should be held until the desired G-level is reached or 3.0 seconds, whichever occurs first.

7.5.7.7. Listen for the first exhalation. It should be short and immediately followed by a quick inhalation. The end of the inhalation may be noted by a sudden grunt sound or a sudden absence of breathing sounds. Total time for the breath ideally should be 0.5 second, but in no case should be longer than one second.

7.5.7.8. Breath sounds should not be heard for 2.5 to 3.0 seconds. If breath sounds are more rapid, average chest pressure is lower and G-tolerance is negatively affected. As G-tolerance is negatively affected, the aircrew will have to work harder at any given G-load. Fatigue during the engagement or especially in subsequent engagements will most likely become apparent. This may be evidenced by even more rapid breathing or breathless, gasping sounds. Observation of the G-load at these times may provide evidence that the individual is apparently working too hard for the G or is unable to maintain the G necessary for the tactical situation.

7.5.7.9. If breath intervals are 4.0 seconds or longer, there is high risk of G-LOC. This is critical when using COMBAT EDGE because it decreases blood flow back to the heart.

7.6. The Aircrew. Survival in the high-G environment can be improved by optimizing our daily habits and activities.

7.6.1. Physical Characteristics. There is no one body type that is immune to G-LOC. Some aircrew are "G naturals", some are not; but ALL of them can significantly increase their G-tolerance. High levels of anaerobic fitness in combination with moderate aerobic capacity are important for all high-G aircrew.

7.6.2. Why Physical Traits Do Not Necessarily Predict G-Tolerance. Evidence suggests that tall individuals are more likely to have a lower resting tolerance but may be able to compensate with an effective anti-G straining maneuver.

7.6.3. How Physical Conditioning Plays an Important Role in G-Tolerance. Anaerobic training has been shown to improve G tolerance in both Air Force and Navy studies. Muscular strength and endurance training has been shown to increase strength, endurance and cockpit mobility during flight in the high-G environment. Moderate amounts of aerobic activity have not demonstrated any degradation in G-tolerance. The cardiovascular fitness improvements from aerobic exercise (aerobic capacity) may enhance the ability to recover from straining maneuvers and shorten the recovery time between engagements and sorties. Attachment 2 to this pamphlet presents a comprehensive program of strength training and aerobic training designed to increase your G-tolerance.

7.6.4. G-Awareness. Application of a timely AGSM is the critical component of high G-tolerance, but G-awareness and "G-discipline" (for high onset rates and sustained G) are the critical components of G-LOC avoidance. G-discipline controls the rate of onset and the level of sustained G; these should be adjusted for multiple human factor variables. Anyone can G-LOC if good G-awareness and discipline are not employed. Pilots who have been "humbled" in an earlier sortie or engagement could have their fangs out further than they realize. They may then apply G-onset rates or sustained G that exceed their capability that day. Ego control and an objective assessment of one's mindset can be a critical factor in the avoidance of a G-LOC. Enhanced situational awareness includes more than just tactical and spatial awareness; it also includes assessing one's own physiological and psychological state throughout the flight. You must develop a subconscious anticipation of impending G-exposure. There is no time to "think about it" with high-G onsets. Using predictive and anticipatory skills will result in proper "G-discipline" in the cockpit. Proper G-discipline, if employed, would have prevented nearly all of our G-LOC fatalities in the CAF. *POINT:* ASSESS BOTH PHYSIOLOGICAL AND PSYCHOLOGICAL STATES DURING HIGH G SORTIES; TO KNOW YOUR LIMITS AND USE GOOD "G-SA".

7.6.5. Technique. Flying technique relates to the previous paragraph as well as to the aircrew's learned air-to-air combat methods. The proper application of G-onset rates and of sustained G is something learned through training and experience. Resist becoming complacent with an aggressive style that disregards these factors. Assess your own technique objectively, and apply the G for the situation. Manhandling the jet is not consistent with optimal "G-SA" and is rarely tactically required if SA is high.

7.6.6. Other Human Factors Affecting Our Ability to Tolerate G-Stress.

7.6.6.1. Fatigue. The ability to perform a maximal AGSM directly correlates to strength and endurance, and these characteristics are related to rest and fatigue. Fatigue from inadequate sleep is different than muscle fatigue. Muscle fatigue severely degrades the AGSM, whereas inadequate sleep degrades alertness and G-awareness. Either way, your G-tolerance is compromised.

7.6.6.2. Heat stress degrades the body's ability to do work and reduces G-tolerance. The combination of dehydration and blood moving to the skin for cooling significantly reduces G-tolerance and work capacity. Studies have shown that with only 3 percent dehydration, G-tolerance time may be reduced up to 50 percent. Always stay well hydrated, especially in hot conditions; increase fluid intake before and during the mission. WHEN HEAT STRESS IS PRESENT, REALIZE YOUR G-TOLERANCE MAY BE REDUCED AND ADJUST YOUR G ONSET RATES AND G-LEVELS ACCORDINGLY!

7.6.6.3. Poor nutrition can affect your performance in the cockpit. It is important to eat regularly two or three times a day. A "proper" diet should emphasize complex carbohydrates (rice, breads, pasta, potatoes, etc.) and less fat and protein than most American diets now contain. Avoid high sugar "coke and candy bar" snacks when you are unable to eat regularly. G-tolerance is reduced if your blood sugar drops below normal ranges. To minimize this potential problem, avoid "substitute" snacks high in sugar content.

7.6.6.4. It has been well documented that alcohol and its hangover effect have a significant negative impact on G-tolerance. Alcohol degrades sleep quality, causes dehydration and salt loss, and depletes body sugar stores. It also tends to dilate blood vessels. All of these factors have a negative effect on your body's ability to tolerate G-stress.

7.6.6.5. Illness or infection also degrades G-tolerance. Although the amount of degradation is unknown, some G-LOC mishaps have occurred just after pilots have returned to flying after being ill. Performance may be reduced after a recent illness, and your body is probably not ready to support the intense effort for an aggressive BFM or ACM sortie. When coming back on status, your body's energy level and muscular strength may be lower and the cardiovascular system may be "detuned" and detrained (slower and less intense response). The use of medication can potentially further degrade your G-tolerance and performance. Together, these effects can reduce your body's ability to tolerate G-stress.

8. G-Protection Equipment.

8.1. Anti-G Garments. The current standard anti-G suit, CSU 13B/P, provides about one G of protection over the relaxed G-tolerance. Its primary purpose is to provide mechanical resistance against which the leg and abdominal muscles can "push", and to reduce the available space for blood to pool, helping to increase blood pressure.

8.2. Life Support Regulations. These regulations require periodic refitting of the anti-G suit and describe the proper fit.

8.3. Future Developments. New G-suit concepts are undergoing development and may be able to afford greater G-protection. Hopefully, they will be incorporated into an integrated life support ensemble in the next generation combat aircraft.

8.4. Pressure Breathing For G (PBG). Pressure breathing for G-protection, COMBAT EDGE, is currently in F-16/15 operational, training, and flight test units. COMBAT EDGE helps increase G-endurance but it does not protect against rapid G-onset. The AGSM is performed the same as if you were flying without COMBAT EDGE. Because of the pressure scheduling and brief delay in onset, COMBAT EDGE DOES NOT PROTECT AGAINST RAPID G ONSET RATES. An anticipatory anti-G straining maneuver is still required prior to the onset of high G to protect against G-LOC with this system. The pressure delivered to the chest cavity provides the equivalent of nearly 3G of protection with aircraft at 9G. The result, sustained G can be tolerated with less fatigue and for greater periods of time. Studies have demonstrated approximately a doubling of G-time tolerance on the centrifuge with assisted positive pressure breathing.

Section C—APPLICATIONS

9. Maximizing G-Tolerance and Preventing G-LOC. Many variables come into play in maximizing G-tolerance and in preventing G-LOC. You, alone, are ultimately responsible for your mental and physical condition and preparedness. How do you stack up? A quick review of the following items before each flight may help you assess your personal capability to fly each time you step to the jet.

9.1. Lifestyle. Regular exercise, good nutrition, and regular sleep allow you to fly at peak condition (peak G capacity).

9.1.1. Participate in a regular exercise program, using both anaerobic (strength) conditioning with moderate aerobic training. The use of strength training with a large anaerobic component (e.g., leg, butt, and abdominal muscles) appears to be the most effective form of muscular training and consequent G tolerance for pilots who fly high-performance aircraft. The use of moderate aerobic training is also recommended since it does not reduce G tolerance and its combination with weight training supplements the endurance criteria required for sustained, prolonged anti-G straining maneuvers.

9.1.2. Practice good nutrition and good sleep discipline. Fly nourished and rested to the maximum extent possible.

9.1.3. Limit the use of alcohol.

9.1.4. Don't fly when ill, fatigued, dehydrated, or while on medications!

9.2. Mission Assessment and Planning. How G proficient are you? Are you rested and fit? Are you well hydrated? Plan your G-employment (onset rates, peak G, sustained G level). Set limits on your G plan of attack.

9.2.1. Consider how long it has been since you were exposed to a high G sortie. If you are coming off a break from high-G missions, plan your G-maneuvering on the assumption that your G-proficiency and G-tolerance will be low. Review the mission, and formulate your own plan for employing high-G maneuvering. Decide what onset rates you will accept, and what G levels you intend to use. Realize you may have to alter your plan inflight if your G-tolerance is not as effective as you had predicted, or if you become fatigued.

9.2.2. Get a good night's rest and a nutritious prebrief meal. Assess your personal fatigue level and adjust your G-plan of attack accordingly.

9.2.3. Drink plenty of water to maintain hydration; don't wait until you feel thirsty.

9.2.4. Although exercise is important, avoid strenuous activity 3 to 4 hours prior to flight due to the temporary reduction of energy stores in the muscles.

9.3. Preflight.

9.3.1. Make sure COMBAT EDGE and the G-suit fits properly and snugly, and that comfort zippers are zipped.

9.3.2. Warm up your muscles in life support, during the walk-around, and in the seat. This should include stretching of the trunk and neck, including rotational movements.

9.3.3. In the cockpit, check G-suit connection and perform ops test.

9.3.4. Review once again the plan: how do you anticipate high-G maneuvering will be employed in this mission, and how do you intend to monitor your G performance? What high-G maneuvers will you NOT accept?

9.4. Inflight.

9.4.1. Performing a G-Awareness Maneuver. This is important in:

9.4.2. Assessing your G-proficiency, and getting back your timing and execution of the anti-G straining maneuver.

9.4.3. Checking G-suit and COMBAT EDGE connections and operation.

9.4.4. Recalibrating how you apply (onset) and control (peak level) aircraft G.

9.4.5. Providing an exposure to the cardiovascular system to activate its reflex response to a drop in blood pressure. A maneuver of 3 to 5G for about 10 seconds is required to fully develop this reflex response.

NOTE: The maneuver is not performed in order to test how many G you can pull that day. Don't G-LOC on the G-awareness turn!

9.4.6. Rehearsing the Anticipation of G-Load. Predict when and how much G will be needed, anticipate the maneuvers, and employ your plan to employ a timely AGSM for G-LOC avoidance.

9.4.7. Performing a Good AGSM Ahead of the G-Onset. Don't wait for the G suit to inflate or for visual symptoms to cue your strain.

9.4.8. Keeping Your "G SA" High. Be aware of how fast and how many G you are commanding. Snatching high G when unprepared may be fatal.

9.4.9. Becoming Fatigued. Remember, as the mission proceeds, you may become fatigued, and your G tolerance may be reduced as your ability to perform an effective straining maneuver is compromised.

9.5. Post Flight. Assess your G-performance. Did you predict your G-tolerance correctly? Did you make appropriate inflight adjustments if needed? From video tape review, did you apply G as expected? Review your AGSM.

9.5.1. Report any physiological problems to your Commander or Supervisor and life support officer. Report any equipment problems to maintenance and life support personnel.

9.5.2. Review the mission on your HUD film; analyze your G-discipline and the effectiveness of your straining maneuver. Did you use more G than you intended? Does your straining maneuver seem to have appropriate pacing? Did you take a quick breath approximately every 3 seconds? Did you blow off too much air? Did you get on top of the G? Did you talk during high-G exposure? Was your G-tolerance lower than expected?

PAUL K. CARLTON, JR.,   Lt General, USAF, MC
Surgeon General

Attachment 1

GLOSSARY OF REFERENCES AND SUPPORTING INFORMATION

References

AFPD 11-4, *Aviation Service*

AFI 11-403, *Aerospace Physiological Training Program*

AFI 11-404, *Centrifuge Training for High-G Aircrew*

Abbreviations

ACM—Air Combat Maneuvers

ACT—Air Combat Training

AGSM—Anti-G Straining Maneuver

BFM—Basic Fighter Maneuvers

CAF—Combat Air Forces

G-LOC—G-Induced Loss of Consciousness

G-SA—Situational Awareness of G

HUD—Head-Up Display

MAJCOM—Major Command

PBG—Pressure Breathing for G

RM—Repetition Maximum

VTR—Video Tape Recorder

Terms

G—Any force that produces an acceleration of 32.2 FPS^2 (FPS = Feet Per Second), which is equivalen to the acceleration produced by earth's gravity.

AFPAM11-419   1 DECEMBER 1999   13

Attachment 2

PHYSICAL FITNESS PROGRAM TO ENHANCE AIRCREW G-TOLERANCE

A2.1. Purpose of the Physical Fitness Program. Flying combat aircraft is a challenging and rewarding career. It takes tremendous desire, knowledge, and skill to be a successful combat aircrew. But winning and losing also depends on your ability to overcome the high G-forces generated by today's modern fighters. This attachment contains specific recommendations for establishment of a personal physical conditioning program to enhance your G-tolerance. Various strength training programs will be described. An outline for a moderate aerobic conditioning program is also provided. This total personal physical conditioning program is designed to increase your strength, endurance and cockpit mobility; in short, to give you the edge in combat.

A2.2. Physical Conditioning To Improve G-Tolerance. Being in good physical condition holds definite advantages for your performance in the cockpit, especially your G-tolerance. Increased muscular strength and endurance allow you to efficiently perform the AGSM with less relative muscular effort, therefore less "mental" effort is required. The AGSM must become an automatic muscle activity, concentrating solely on performance of the AGSM will erode your attention to such tasks as good BFM. During fighter maneuvers, high G-forces may be sustained for a short duration requiring performance of the AGSM. This maneuver expends energy at a rate similar to sprinting or weight lifting. Researchers have shown that high intensity strength training increases your ability to withstand high G-forces for a longer duration with less fatigue. Additionally, moderate aerobic training has been shown to decrease the recovery time needed between centrifuge training runs. Therefore, a combination of both training programs is recommended for maximum G-tolerance and performance.

A2.3. High Intensity Muscular Training.

   A2.3.1. Program Requirements. A well balanced muscular conditioning program must consist of two parts: strength AND muscular endurance. It requires a good deal of strength to restrict blood flow to the skeletal muscles. It is also important to maintain the muscular contraction for the period of time spent under the G-load, as well as repeating the contraction time and time again with equal effectiveness. This aspect of the strain requires muscular endurance. Since it is critical to have both strength and muscular endurance, a well balanced program will incorporate both.

   A2.3.2. Strength and Power Workout. As with any type of training, strength training workouts should begin with a brief 5 to 10 minute warm-up period. This warm-up should include stretching and slow jogging in place or brisk walking. Rest periods of 1.5 to 2 minutes should occur between sets and between exercises. The weight that is to be used should be calculated from a "one repetition maximum (RM)" lift. One RM is the most weight that can be lifted once in a slow and controlled fashion for a given exercise. This amount should be established for each exercise that will be done during the workout. Once this amount is measured, multiply it by 0.8 to find 80 percent of the 1 RM. This weight is the training weight for any given strength exercise. Each set should consist of six repetitions. One repetition is when the weight has been lifted then returned to the starting position. For example, in a bench press, one repetition would be lifting the weight until the elbows are straight (but not locked), and lowering it back to the chest. After six repetitions are completed, the rest period begins. Three sets of the same exercise should be accomplished during each workout. All three sets should be completed before another exercise is begun. Weights should always be lifted slowly and deliberately, being lowered slower than they were lifted. Some find it helpful to lift to the count of 2 and lower to the count of 4.

A2.3.3. Endurance Resistance Workout. A 5 to 10 minute warm-up similar to that described for the strength workout should precede the endurance workout as well. The rest periods for endurance workouts should be 1 to 1.5 minutes. Once again, the 1 RM for each exercise must be determined. The 1 RM is multiplied by 0.6 to find the weight to be used for training. Three sets of each exercise are to be completed with each set containing twelve to fifteen repetitions. If desired, endurance workouts may combine two separate exercises that are training the same muscle group. In this case, the individual would complete a set of exercise A, rest 1 to 2 minutes, then complete a set of exercise B, rest, return to exercise A, and so on until three sets of both A and B are completed.

A2.3.4. Combination Program. To simply combine both programs for muscular strength and endurance training, use the 1 RM times 0.7 (70 percent). This method will give both strength and endurance gains without neglecting either. The rest period for this workout program should be 1.5 to 2 minutes. There should be three sets of eight repetitions.

A2.3.5. Reevaluating 1 RM. 1 RM should be rechecked after four weeks of training in order to increase the weight being lifted as strength increases. Another way to increase weight is to add 5 to 10 percent more weight once a person is able to complete an additional two repetitions at the current weight for the strength program or an additional four repetitions for the endurance program after completing the last set.

A2.3.6. High Intensity Circuit Training. This strength and endurance training program is a high intensity activity. Exercise sets are performed one muscle group after another. 60% of 1 RM is used, exercises are performed at 12 - 15 repetitions for each exercise. A sample exercise rotation would be as follows: 12 repetitions of quadriceps, hamstring, calf, butt then abdominal group. A two minute rest period then begin again. Three times through this circuit constitutes one exercise session. Once you can complete the exercise session with 12 repetitions each exercise work to reduce the rest period to 30 seconds. Cardiovascular training must be accomplished along with this workout.

A2.4. Specific Exercises. Regardless of the type of program (strength, endurance or combination) being followed, it should include at least two exercises for each of the muscle groups listed below. For specific exercises you should consult the local Aerospace Physiologist, human performance team member, Exercise Physiologist, or fitness center instructor.

A2.4.1. Legs: leg extensions, leg curls, quarter or half squats, lunges, and toe raises.

A2.4.2. Chest: wide and narrow grip bench press, incline bench press, fly's, and pullovers.

A2.4.3. Arms: biceps curls, triceps extensions, and dips.

A2.4.4. Abdominals: stomach crunches, cross crunches, leg drops, reverse stomach crunches, and flutter kicks

A2.4.5. Neck and shoulder: shrugs and upright rows.

A2.4.6. Butt: flutter kicks and, quarter squats.

A2.5. Aerobic Conditioning. In addition to weight training, it is important to participate in some type of moderate aerobic training. Aerobic conditioning develops a stronger, more efficient heart and increases the blood supply to the working muscles which will significantly reduce recovery time between engagements and sorties. In a combat situation, more aerobically fit individuals will better tolerate the multiple sorties per day schedule. The recommended program consists of 20 to 60 minutes of activity, three times per week, at an intensity that elicits a heart rate in the target zone for that individual. [Target heart rate zone = (220-age) x 0.6-0.9] Another way to measure intensity is a talk test: while performing any aerobic activity the participant should be able to converse with relative ease and breathing should not be labored. The type of activities which lend themselves to aerobic conditioning include: running, jogging, swimming, cycling, rowing, stair climbers and aerobic dance. Activities not included are: racquetball, basketball, golf and tennis [these are good fitness activities but they do not rely exclusively on the aerobic system for the energy required].

Index